SYSTEMS BIOLOGY OF CLOSTRIDIUM

SYSTEMS BIOLOGY OF
CLOSTRIDIUM

Editor

Peter Dürre
University of Ulm, Germany

ICP
Imperial College Press

Published by

Imperial College Press
57 Shelton Street
Covent Garden
London WC2H 9HE

Distributed by

World Scientific Publishing Co. Pte. Ltd.
5 Toh Tuck Link, Singapore 596224
USA office: 27 Warren Street, Suite 401-402, Hackensack, NJ 07601
UK office: 57 Shelton Street, Covent Garden, London WC2H 9HE

Library of Congress Control Number: 2014937637

British Library Cataloguing-in-Publication Data
A catalogue record for this book is available from the British Library.

SYSTEMS BIOLOGY OF CLOSTRIDIUM

ISBN 978-1-78326-440-7

Printed in Singapore by FuIsland Offset Printing (S) Pte Ltd

TABLE OF CONTENTS

PREFACE

Systems biology is a rapidly evolving scientific discipline, which aims to elucidate, understand and predict the behaviour of complex metabolic and controlling networks within a single bacterial cell or even interactions in microbial communities (at the level of prokaryotes). Results from experiments are coupled to mathematical modelling, which in turn will generate predictions that can be tested experimentally, thus leading to constant improvement of the model in an iterative approach. This book provides a comprehensive overview of this interdisciplinary field for the genus *Clostridium*. This is a large and heterogenous group of bacteria, with great potential for industrial application. The best known species is certainly *Clostridium acetobutylicum*, which first performs a typical acidogenic fermentation (butyrate and acetate), before switching to the production of solvents (butanol and acetone). It was this feature that led to the enormous industrial use of the acetone-butanol-ethanol (ABE) fermentation during the first half of the 20th century and its revival in countries such as Brazil and China in recent years. The interest in biotechnological processes for butanol production has risen dramatically as butanol is not only an important bulk chemical but also a superior biofuel to ethanol and other fossil resources processed from crude oil. The underlying metabolic pathways and the controlling network are complex, rendering a systems biology approach for elucidation of the respective interactions promising.

The importance of the clostridia for biotechnology is also emphasized by the steadily growing interest and the increasing number of participants at the biannual international *Clostridium* conferences. The first conference was held in 1990 in Salisbury, UK, followed by symposia in Blacksburg, USA, Evanston, USA, Ulm, Germany, Toulouse, France, Champaign/Urbana, USA, Rostock, Germany, Edinburgh, UK, Houston, USA, Wageningen, the Netherlands, San Diego, USA, and Nottingham, UK. The 2014 conference is to take place in China. Interested scientists and advanced students are invited to check the respective websites (e.g. http://www.clostridia.net) for further information and future participation. This book should not only serve as a reference for researchers in this field, but also attract graduate students and advanced undergraduates with an interest in a rapidly emerging field, providing significant tools and results for further industrial exploitation.

Most of the authors are members of a European programme, funding basic research on the systems biology of microorganisms (SysMO, http://www.sysmo.net). This programme was initiated in 2007 under the auspices of Austria, Germany, the Netherlands, Norway, the UK, and Spain and come to an end in 2013. The clostridial project in this programme was designated COSMIC (*Clostridium acetobutylicum* systems microbiology). It was this financial contribution that led to most of the results presented in this book. All COSMIC participants thus gratefully acknowledge the support of BBSRC, BMBF and NWO.

Ulm, 2013
Peter Dürre

CHAPTER 1

METABOLIC AND REGULATORY NETWORKS IN *CLOSTRIDIUM ACETOBUTYLICUM*

P. GÖTZ

Bioprocess Engineering
Beuth University of Applied Sciences Berlin
Seestrasse 64, 13347 Berlin, Germany

M. REUSS

Center for Systems Biology
Nobelstrasse 64, Stuttgart, Germany

A. ARNDT, P. DÜRRE

Institute of Microbiology and Biotechnology
University of Ulm
Albert-Einstein-Allee 11, 89069 Ulm, Germany

Solvent production from clostridia began 100 years ago and evolved into one of the largest global biotechnological processes in the first half of the 20th century. Despite its rather low level of productivity, this process remained feasible until it was replaced by more economic crude oil-based production. In reviving the process, sparked by rising crude oil prices, optimization of yield and productivity is essential. Modern scientific tools, integrating methods from molecular biology to mathematical modeling and process control, are employed to gain a thorough understanding of interactions during the process at molecular, cellular and bioreactor levels. A systematic approach with which to implement this interdisciplinary research is to construct a metabolic network from the sequenced genome. Such a genome-wide metabolic network is the basis for the set-up of mathematical equations as well as for studying regulatory interactions between its components. This first step, accompanied by suggestions for further work, is presented in this chapter.

1.1 Introduction

With the recurring interest in the biotechnological production of solvents due to increasing crude oil prices over the past two decades, the detailed examination of the biphasic fermentation metabolism of *Clostridium acetobutylicum* has moved back into focus. Today, it is generally understood how the enzymatic reactions leading to acetic and butyric acid formation function during the exponential phase of growth, how these acids are converted into the solvents acetone and butanol with entry into the stationary phase, as well as the functioning of the accompanying sporulation later in the stationary growth phase. Furthermore, the genes encoding the key enzymes involved have been identified and well characterized. In contrast, the regulatory aspects of this distinctive metabolism are still largely unknown but, it goes without saying that, for the development of an efficient industrial solvent production process, a detailed knowledge of the regulation of this metabolic shift is indispensable.

Quantification of knowledge about living systems is ultimately achieved through mathematical modeling. The joint efforts of advanced disciplines in tackling this complex task is known as systems biology.

Before defining a methodological framework for undertaking this task, the purpose of this effort must be stated. Although the beauty of an *in silico* cell may be a reason in itself for creating it, this would not justify the time and money spent on such an activity. In our case, the increase of butanol yield, including high productivity and high selectivity towards this product are the ultimate goals. Systems biology and the resulting quantification of microbial behavior in a predictive model will enable these goals to be reached via metabolic engineering or synthetic biology. In metabolic engineering, the mathematical model is used to identify targets for genetic modification as well as to select environmental conditions in a controlled bioreactor in order to achieve the desired process characteristics. For synthetic biology, metabolic subnetworks (e.g. single pathways) may be isolated from the overall network model together with their regulatory interfaces, and may be recombined into new functional units. These can then be the templates for the creation of new genetic information, leading to a synthetic cell.

After defining the purpose of applying the systems biology toolbox, the cyclic iterative workflow between the participating experimental and modeling groups must be established. It is mostly useless to attempt to find meaning in deliberately collected data. From the beginning, there should be a systematic approach, model-based and using hypotheses to advance step by step towards the ultimate mathematical model which collects and quantifies all experimental observations. Although the mathematical and computational tools for quantification of biochemical reaction systems are well developed, there is still no standard method for representing metabolic and regulatory networks in cells. For changes in and fluxes between metabolite pools, mass balances are usually set up. The resulting system of ordinary differential equations (ODEs) implies large numbers of molecules which are homogeneously distributed in the cell volume. For prokaryotic cells, this assumption is acceptable, higher cells may require more detail in modeling with respect to spatial distribution of proteins and metabolites. Selecting a method for representation of regulatory interactions is more controversial. Signal molecules and especially binding sites in a cell are quantified by small integer numbers. Although this would not allow the application of mass balance-based ODEs for representation of these compounds and binding sites in a cell, the assumption of a large population of identical cells in a reactor, lumping their volume into a total cell volume makes the application of ODEs acceptable. Therefore a first roadmap for a systems biology approach to describe the complex behavior of a microorganism would start from the sequenced genome with the following steps:

1. Create a genome-wide metabolic network, containing all possible pathways.
2. Split a genome-wide network into sub-networks for feasible experimental investigation.
3. Create a mathematical model by setting up metabolite mass balances.
4. Start a cyclic workflow of model verification/validation and experimental investigation.

1.2 Roadmap to modeling

The first step in the above roadmap is usually approached by starting from gene annotation on the genome, selecting enzymes with known function, connecting known pathways via common intermediates, postulating missing intermediates and related enzymes, verifying the function of postulated enzymes by comparison of protein sequences with known enzymes from other microorganisms, thereby completing the network. For visualization and annotation of the information related to the network components, the software package Insilico Discovery^{TM}, Version 3.0.0 (Insilico Biotechnology AG, Stuttgart) was used. The result is a network consisting of 298 intracellular metabolites which are stoichiometrically balanced (intracellular water and hydrogen are not balanced) and 17 unbalanced extracellular metabolites. The metabolites are connected by 512 transforming steps (29 membrane transport and 483 biochemical reaction steps). The network is organized into 63 pathways. Visualization of this network is shown in Fig. 1.1.

In Fig. 1.1, dots represent metabolites (stoichiometrically balanced: grey, unbalanced, e.g. H_2O: red) and squares represent transformers (transport steps: yellow, reaction steps/enzymes: red). For ease of visibility, not all connections (black lines) between the metabolites are shown. To give a better overview, the main purpose of different sections of the network is highlighted by adding frames.

Metabolite annotation of this network allows accommodation of the following information:

- Identifier section
  Identifier / Trivial name / Systematic name / Compartment / CAS number / Kegg ID / PubChem
- Properties section
  Elemental composition / Charge / Molecular weight / molecular structure (SMILES) / Balanced
- Connections section
  List of transformers which are connected to this metabolite
- Concentration section
  Concentration values / Standard deviations

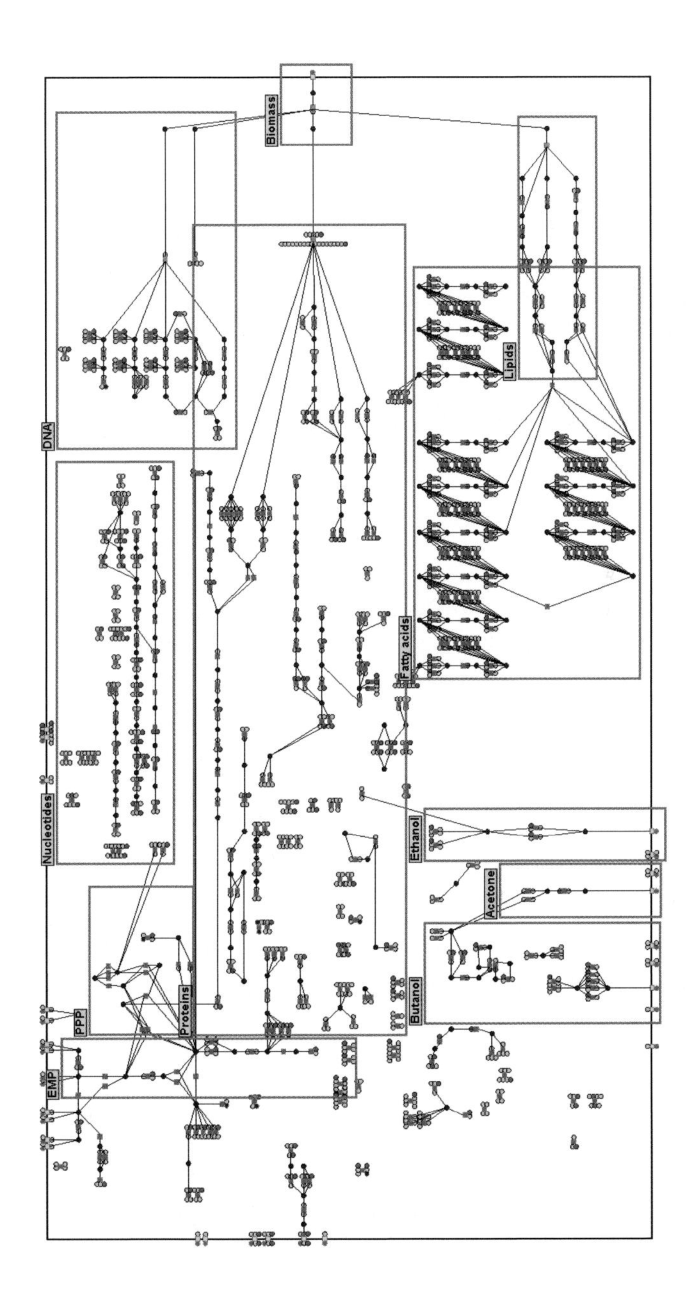

Fig. 1.1: Metabolic network of *Clostridium acetobutylicum* ATCC 824, visualized by Insilico Discovery™.

- Initial value section
  Slider for selecting the initial value of the metabolite for network simulations
- Comments section (Comments on related publications, assumptions during network creation, etc.)

An example of the corresponding software window is shown in Fig. 1.2.

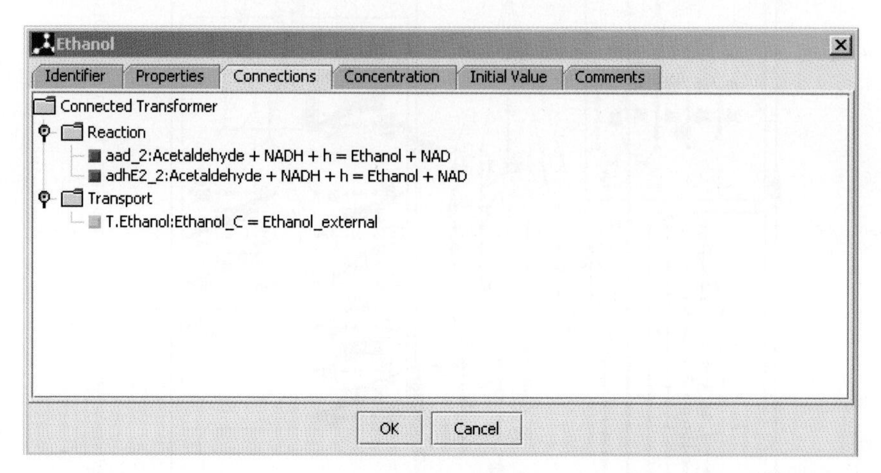

Fig. 1.2: Metabolite annotation window (Insilico Discovery™).

Transformer (e.g. enzyme catalyzing a biochemical reaction) annotation of this network allows accommodation of the following information:

- Identifier section
  Identifier / Trivial name / Systematic name / Equation / Compartment / Pathway / Kegg ID / Gene ID
- Kinetics section
  Reversibility of reaction / Modulation of reaction by activators or inhibitors / Elasticities (Sliders for substrates, products, activators and inhibitors) / Capacity / Catalysis / Trajectory
- Steady-state section
  Value for calculated steady-state flux
- Comments section
  Comments and related literature / Curation (manual / automatic)

An example is shown in Fig. 1.3.

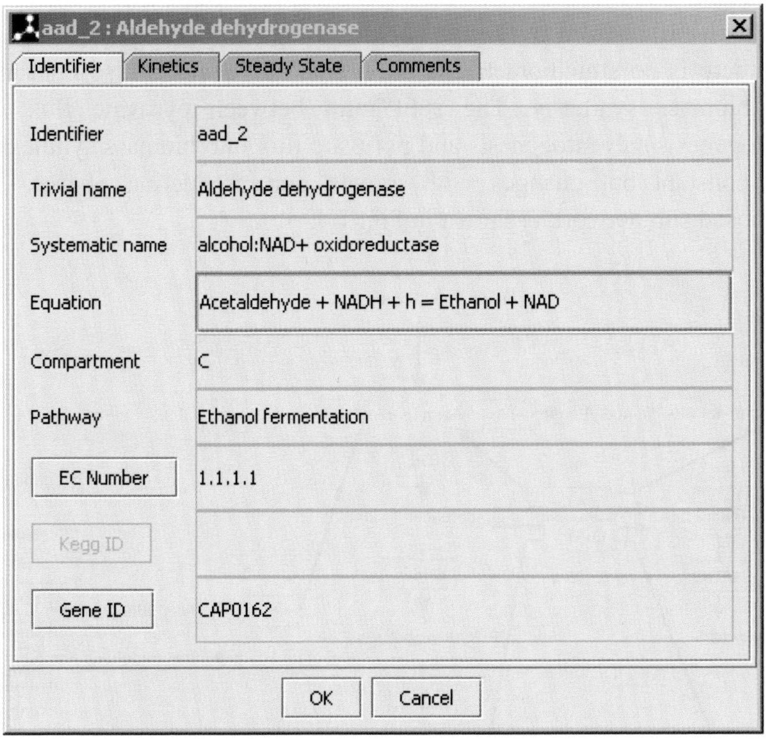

Fig. 1.3: Transformer annotation window (Insilico Discovery^{TM}).

Structural information and annotation were exported into SBML format and are available on request via the SysMO database SysMO-SEEK (seek.sysmo-db.org). Other reports on the creation of metabolic networks for *C. acetobutylicum* are published[1,2,3], but detailed comparison of the networks is not possible, since these networks are not available in a standardized format. The networks are of similar size though, 552 reactions and 422 metabolites[1] and 502 reactions and 479 metabolites[3].

Since the experimental effort for flux analysis of the complete network is huge, the proposed second step is now the isolation of a sub-network. Usually, the central carbon metabolism is chosen, initially we further simplify this to an acidogenic/solventogenic metabolism with a carbon influx from pyruvate. This influx is related to glucose uptake, yet there is no stoichiometric relation since pyruvate is also consumed for biomass synthesis. The split ratio between pyruvate flux into acidogenesis/solventogenesis and pyruvate flux into biomass synthesis is not constant but changes with growth and production phases. The proposed sub-network is shown in Fig. 1.4.

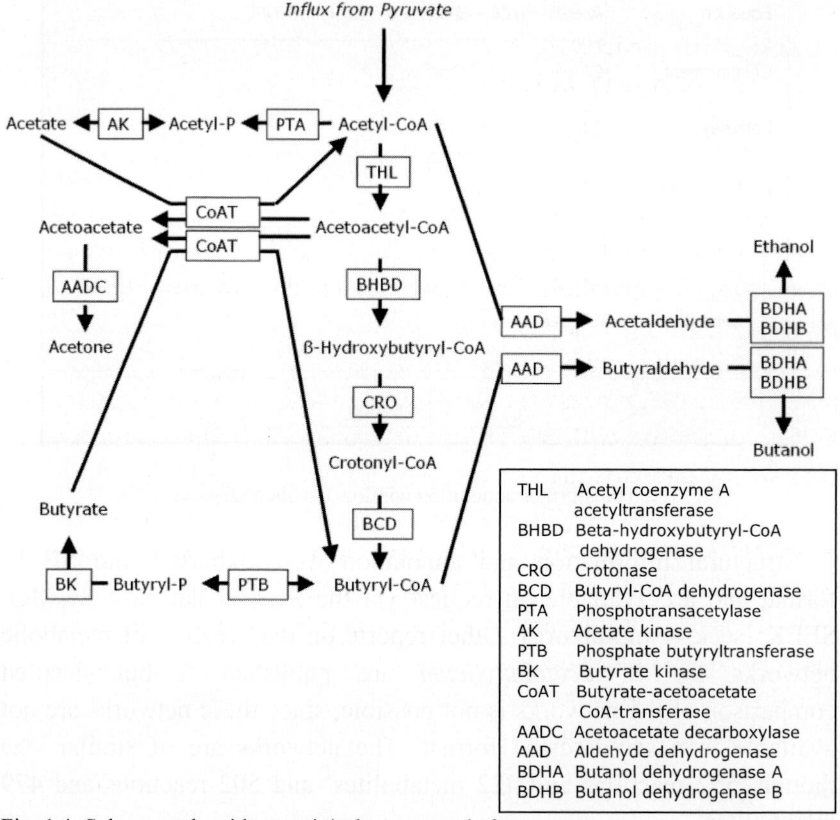

Fig. 1.4: Sub-network acidogenesis/solventogenesis from pyruvate.

Due to the temporal sequence of acidogenesis and solventogenesis, this network may be further simplified into two networks which are exclusively active, regulation toggling between these two modes of activities. During acidogenesis, only reactions downward along the center and directed to the left within the network shown in Fig. 1.4 are active. During solventogenesis, acid uptake from the left side coupled to acetone formation and conversion reactions into the other solvents to the right are active, while pyruvate influx may still be present.

Step three in the proposed workflow towards an *in silico* cell is the set-up of mass balances for all components within the sub-network. Starting with a description of a batch process, the corresponding reactor working volume V_R of the batch reactor is a two-phase system, consisting of particles (cells) and the extracellular bulk liquid. The respective volumes in the batch reactor are defined as:

$$V_R = V_C + V_L = \text{const.} \tag{1.1}$$

The assumption of a constant reactor volume, being the definition of a batch process, may be questioned in this context for a real fermentation experiment. Evaporation from the reactor due to gas flow, water production during central carbon metabolism, addition of correction fluids, etc. may change the volume. Therefore, a constant volume for a real fermentation process is a simplification to begin with. Nevertheless, a strict derivation will be performed to make all simplifications clear. A first assumption is the dry weight of cells being constant (commonly used first estimate: 300 $g_{DW}/liter_{cells}$). For the mass balance of component A in the extracellular liquid, there is no biochemical reaction assumed, changes in extracellular concentrations are only caused by fluxes to and from the cells. For practical reasons, the specific fluxes v_A are related to the working volume, since they may change direction during the process. The direction of v_A is defined as positive for a flux from the cell into the liquid. From the mass balance, the equation of change for $c_{A,ex}$ is derived:

$$\frac{dc_{A,ex}}{dt} = \frac{\frac{\mu \cdot c_X}{\rho_X} \cdot c_{A,ex} + \varphi_A}{\left(1 - \frac{c_X}{\rho_X}\right)} \tag{1.2}$$

The interpretation for the left-most term in the numerator is a rise in extracellular concentration of A caused by a decreasing liquid volume due to the increase of cell volume by growth. The denominator accounts for the reactor volume-based flux causing a change in the liquid-based extracellular concentration of A. For example for zero growth, the concentration change in the liquid volume is greater compared to the total flux which is related to the greater reactor volume. The influence of these two terms is usually neglected, which is acceptable for low biomass concentrations. The equation of change for the intracellular compound A is:

$$\frac{dc_{A,in}}{dt} = r_{A,in} - \frac{\rho_X}{c_X} \cdot \varphi_A - \mu \cdot c_{A,in} \tag{1.3}$$

Production of A by intracellular reaction rate $r_{A,in}$, transport flux from the cell and dilution of the intracellular component by growth are the terms on the right side, respectively. Simplifying the equations by assuming rapid equilibrium between the intracellular and extracellular phase allows the transport equation to be neglected. Fluxes from/into the extracellular component pool can be replaced by the corresponding reaction rate. For conversion of the biochemical rate of catalysis within the cell $r_{A,in}$ and the rate of conversion r_A related to the reactor volume, the following equation holds:

$$r_A = r_{A,in} \cdot \frac{V_C}{V_R} = r_{A,in} \cdot \frac{c_X}{\rho_X} \tag{1.4}$$

Assuming rapid equilibrium, neglecting the two terms in Eq. (1.2) as proposed (relatively low biomass concentration in reactor, $c_X << \kappa_X$) and defining the rates of conversion related to the reactor volume, Eq. (1.2) for the extracellular compound simplifies to:

$$\frac{dc_{A,ex}}{dt} = r_A \tag{1.5}$$

Accordingly, Eq. (1.3) simplifies to:

$$\frac{dc_{A,in}}{dt} = \frac{\rho_X}{c_X} \cdot r_A - \mu \cdot c_{A,in} \tag{1.6}$$

The sub-network in Fig. 1.4 is the structural basis for the set-up of the equations of change for the respective metabolites. Simplifying the sub-network by lumping purely consecutive reactions into one reaction step, a system of differential equations (Fig. 1.5) is created. The contribution of the "dilution by growth" related term μAc_i within the equations for intracellular compounds is often neglected, since it is small compared to the fluxes through the intracellular metabolite pools. Acetyl-CoA formation from pyruvate is expressed by the "influx" term. As previously stated, this influx cannot be calculated from glucose uptake due to

1: Acetyl-CoA, C2, intracellular, mmol/l_C	$\frac{dc_1}{dt}=\frac{\rho_X}{c_X}(Influx - r_{1\_2} + r_{2\_1} - r_{23\_17} - r_{11\_3} - r_{1\_9}) - \mu c_1$
2: Acetate, C2, extracellular, mmol/l_R	$\frac{dc_2}{dt}=r_{1\_2}-r_{2\_1}-r_{23\_17}$
3: Acetoacetyl-CoA, C4, intracellular, mmol/ l_C	$\frac{dc_3}{dt}=\frac{\rho_X}{c_X}\left(\frac{r_{11\_3}}{2} - r_{3\_7} - r_{3\_4}\right) - \mu c_3$
4: Butyryl-CoA, C4, intracellular, mmol/ l_C	$\frac{dc_4}{dt}=\frac{\rho_X}{c_X}\left(r_{3\_4}+r_{63\_47}-r_{4\_5}+r_{5\_4}-r_{4\_10}\right)-\mu c_4$
5: Butyryl-Phosphate, C4, intracellular, mmol/ l_C	$\frac{dc_5}{dt}=\frac{\rho_X}{c_X}\left(- r_{5\_4} - r_{4\_5} - r_{6\_5} - r_{5\_6}\right) - \mu c_5$
6: Butyrate, C4, extracellular, mmol/l_R	$\frac{dc_6}{dt}=r_{5\_6}-r_{6\_5}-r_{63\_47}$
7: Acetoacetate, C4, intracellular, mmol/ l_C	$\frac{dc_7}{dt}=\frac{\rho_X}{c_X}\left(r_{3\_7}-r_{7\_8}\right)-\mu c_7$
8: Acetone, C3, extracellular, mmol/l_R	$\frac{dc_8}{dt}=r_{7\_8}$
9: Ethanol, C2, extracellular, mmol/l_R	$\frac{dc_9}{dt}=r_{1\_9}$
10: Butanol, C4, extracellular, mmol/l_R	$\frac{dc_{10}}{dt}=r_{4\_10}$

Fig. 1.5: Simplified mathematical model for acidogenesis/solventogenesis from pyruvate.

changing yields. From a carbon balance perspective, this influx may be calculated by summing up the fluxes into/from the extracellular compounds (acetate, butyrate, acetone, ethanol, and butanol) and taking into account the stoichiometrically coupled CO_2 evolution.

This may be incorporated directly into the system of equations, replacing the influx term by a combination of the other fluxes via carbon balance. Another approach is to create a calculated data set via carbon balance for the influx and using it for further analysis of the system. The mathematical model in Fig. 1.5 can be used to calculate the fluxes/ reaction rates during fermentation from a set of product concentration time series data. For batch fermentation, this constitutes a metabolic flux analysis (MFA) over the dynamic course of this process. A necessary condition for this computation is neglecting the changes of the intracellular metabolites related to the fluxes through the respective metabolite pools (setting the left sides of Eqs (1.1), (1.3), (1.4), (1.5) and (1.7) to zero). The result of this MFA is only descriptive, since it connects the different extracellular metabolite data time series by intracellular fluxes. An interesting result from this analysis would be the determination of the fraction of pyruvate influx (from glucose) going directly into solvents, since maximizing this fraction would be a desired feature for an efficient solvent production process. Using the model structure in Fig. 1.5 for implementation of a predictive model, mathematical expressions for the rates of conversion must be formulated. These rates usually have a maximum value, depending on the enzyme (catalyst) concentration, which is then modified by the kinetic behavior of the enzyme. The kinetic behavior in turn depends on substrate, inhibitor and activator concentrations – mathematical expressions can be chosen according to the reaction mechanism. For example, for the reaction from metabolite i to metabolite j, assuming Michaelis–Menten reaction kinetics regarding substrate concentration c_i and inhibition by metabolite concentration c_k, the expression for the reaction rate could be:

$$r_{i,j}(t) = k_{i,j} \cdot c_{Enzyme,i,j}(t) \cdot \frac{c_i(t)}{K_{M,i,j} + c_i(t)} \cdot \frac{K_{I,i,j,k}}{K_{I,i,j,k} + c_k(t)} \qquad (1.7)$$

Model parameters are $k_{i,j}$ (reaction rate constant), $K_{M,i,j}$ (Michaelis–Menten constant) and $K_{I,i,j,k}$ (inhibition constant of inhibitor k for the reaction from i to j). The time dependency of the concentrations on the right side of Eq. (1.7) is explicitly stated to underline the difference to constant model parameters. The simplest case for an enzyme reaction kinetic expression is the Michaelis–Menten type. Further influences and mechanisms can be inferred from *in vitro* and *in vivo* data to expand this expression. For modeling of *Clostridium acetobutylicum*, a recent publication reviews the knowledge on enzymatic conversions[4]. Kinetic rate expression structures and first estimates for the model parameters can be taken from this and other related publications. Although this may allow collection of all necessary information on metabolites which are directly influencing the enzymatic reaction, the enzyme concentration is still unknown in this equation. The amount of intracellular enzyme is mainly governed by enzyme expression, enzyme degradation and dilution by growth. Enzyme expression in turn is governed by regulation, a complex combination of processes including, for example, the induction and/or repression of transcription on a genomic level. Linking cause and effect when changes of enzyme concentration in the cell occur requires knowledge of the key molecular components involved in this process. Regulatory proteins, binding sites, phosphorylation in signaling cascades, may all play a role in the resulting regulatory network, which is coupled to the metabolic network. Unraveling this regulatory network to enable a predictive model for mechanistic description of changes in enzyme concentrations in central carbon metabolism is a major challenge. So far, only a few of the key players in this regulatory network have been identified.

1.3 Regulators of solventogenesis and sporulation in *C. acetobutylicum*

The first experimental evidence for a regulatory link connecting the metabolic networks of solventogenesis and sporulation in *C. acetobutylicum* was only obtained in 2000. Ravagnani *et al.* discovered that the master regulator of sporulation, Spo0A, also directly controls the shift from acid to solvent production[5]. Their conclusions were based on *in vitro* gel retardation experiments using the *adc* gene promoter region as a probe (*adc* encodes acetoacetate decarboxylase which is required for acetone formation) and Spo0A from either *Bacillus subtilis* or *C. beijerinckii*. Spo0A-binding motifs (0A boxes) were identified and targeted mutations showed dramatic effects on binding (no binding after destruction of the 0A boxes, enhanced binding after precise conversion to the consensus sequence). For *C. beijerinckii*, reporter gene constructs and insertional inactivation of the *spo0A* gene validated the conclusions. Further confirmation for *C. acetobutylicum* was provided by construction and analysis of an *spo0A* inactivation mutant[6]. In all these experiments, formation of acetone and butanol was similarly affected. This is in accordance with the presence of 0A boxes upstream of the *adc* and *sol* operons (the latter encoding AdhE, a bifunctional butyraldehyde/butanol dehydrogenase, and Ctf, encoding an acetoacetyl-CoA:acetate/butyrate coenzyme A transferase). Usually, butanol and acetone are produced in a ratio of 2:1. However, conditions are known that lead to dramatic changes. When using whey as a substrate, a hundred-fold more butanol was synthesized[7] and the combination of sugar and reduced carbon sources such as glycerol leads to a so-called alcohologenic fermentation with butanol and ethanol as the only solvents formed[8]. Under the latter conditions, butanol production is catalyzed by AdhE2, a second bifunctional butyraldehyde/butanol dehydrogenase present in *C. acetobutylicum*[9]. This means that additional regulators must be involved in solventogenesis. Also, knock-out of 0A boxes upstream of the *adc* gene of *C. acetobutylicum* did not completely abolish the regulatory pattern[5]. An experimental approach to find and identify such additional regulators was based on DNA affinity chromatography. Biotinylated DNA fragments carrying the promoter regions of the *adc* and *sol* operons were bound to streptavidin-coupled magnetic beads and incubated with cell extracts of *C. acetobutylicum*. Magnetic separation and washing away of unbound molecules led to DNA fragments carrying regulatory proteins. In the case of the *adc* promoter, a novel transcription factor (AdcR) could be identified[10], whereas in the case of the *sol* promoter two known global regulators were found, namely CcpA and CodY[11]. With respect to systems biology and the computational generation of regulatory network models, it seems crucial for model quality to provide the modeler with as much experimental data as possible. For this purpose, a thorough characterization of the identified regulators in terms of binding sites and co-factors and/or signaling molecules is necessary and, in fact, current studies are directed towards the molecular characterization of these clostridial regulatory proteins, their function as well as their interplay in regulation of solvent formation.

Regarding the characterization of the novel regulator AdcR, a detailed sequence analysis identified a helix-turn-helix DNA binding domain at the N-terminal end of the protein. Interestingly, AdcR also exhibits a putative transmembrane domain at its C-terminal end which could indicate a potential anchorage of the protein in the membrane of the cell[12]. The AdcR encoding gene, *adcR*, is located on the megaplasmid pSOL in close proximity to the second bifunctional aldehyde-/alcohol dehydrogenase gene *adhE2* and expression analysis revealed that *adcR* is transcribed together with the downstream-located gene *adcS* with maximal expression during the acidogenic phase of growth[13]. Although AdcS did not bind to the promoter regions of either the *sol* operon or the *adc* gene in the performed DNA affinity chromatography experiments, the presence of a single *adcR/adcS* transcript indicates that this protein also plays an important role in transcriptional regulation of genes involved in solventogenesis. In fact, the simultaneous incubation of both proteins in electromobility shift experiments resulted in a diminished retardation of the *adc* promoter region, dependent on AdcS concentration, hinting at a counteractive function of this protein to AdcR activity and, therefore, AdcS might be fine-tuning the expression of the affected gene(s)[14]. When AdcR is being overproduced in the cell, the concentration of solvents decreases significantly, indicating a repressing role for the protein in both *adc* and *sol* operon expression. It becomes more and more apparent that a complicated interplay of AdcR and AdcS in regulation of solvent formation most likely exists and further experimentation to elucidate their role and to gain more accurate data for modeling is necessary.

In contrast to AdcR and AdcS, CcpA is a well known pleiotropic control protein in Gram-positive bacteria. In the past, CcpA was mostly known for its role in catabolite repression in low GC Gram-positive bacteria. In this context, CcpA binds to its serine-phosphorylated co-repressor HPr and, together, this complex is able to bind to so-called *cre*- (catabolite responsive element) sequences on the DNA. In fact, it has been shown that the affinity of CcpA to DNA, when not bound to HPr, is significantly decreased. Dependent on the localization of the *cre*-sequences with respect to the transcriptional start, CcpA exhibits either an activating or a repressing function to the expression of the respective gene(s). Various publications report on the glucose-mediated repression of genes involved in the metabolization of alternative carbon sources in *C. acetobutylicum*, showing that catabolite repression mediated by CcpA is also an important regulatory mechanism in this organism[15,16,17,18,19]. The binding of CcpA to the promoter region of the *sol* operon allocates a regulatory function of this protein in *C. acetobutylicum* as well, i.e. the involvement in regulation of solvent formation and, in fact, a potential *cre*-sequence (AATTATACGTTTACA), which shows similarity to the known consensus *cre*-sequence TGTAAGCGTTAACA, has been identified[20]. *In vitro* binding studies indicate a high dependency on phosphorylated HPr for a specific binding of clostridial CcpA[20], a fact already known for CcpA proteins of other Gram-positive organisms. Studying the direct influence of CcpA on *sol* operon expression has turned out to be complicated since, until recently, it has been proven rather difficult to inactivate or delete the *ccpA* gene and, as a matter of fact, it was only in 2010 that Ren *et al.* reported on the disruption of *ccpA* by insertion of a group II intron[18]. Characterization of a *ccpA*-negative mutant in terms of the regulatory effects on solvent formation, however, should provide important data for modeling approaches. A further aspect that has to be taken into account is the dependency of CcpA on phosphorylated HPr and consequently on fructose 1,6-bisphosphate. The intracellular concentration of co-factors is highly valuable information for computational modeling and, therefore, establishment of proper methodology for the determination of these metabolites is currently tackled within the systems biology approach.

Like CcpA, CodY is also a well-known global regulatory protein in Gram-positive bacteria regulating the expression of a variety of different genes, e.g. genes involved in amino acid biosynthesis, transport or sporulation mechanisms[21]. The binding of CodY to the promoter region of the *sol* operon in *C. acetobutylicum* indicates an important role in the regulation of solventogenesis in this organism. Similar to other bacteria, the affinity of the clostridial CodY to DNA seems also to be dependent on the co-factor GTP as shown in *in vitro* gel shift experiments[20]. Using DNA footprinting analysis, the sequence AATATACTGATAATT was identified as functioning as a binding site for CodY in the *sol* promoter region (CodY box)[20]. Since CodY has already been described in several Gram-positive bacteria, it is not surprising that the clostridial CodY box shows great similarity to the well-known consensus sequence AATTTTCWGAAAATT[21,22,23]. Interestingly, the inactivation of *codY* leads to no altered growth behavior, although both the final butanol concentration and the final acetone concentration of mutant cultures is significantly increased, indicating a repressing function of CodY in solventogenesis. Moreover, while the formation of solvents in wild-type cultures takes place in the mid- to late-stationary phase at a pH value of about 4.5, the onset of butanol and acetone formation in mutant cultures already occurs in the late-exponential and early stationary growth phase and, therefore, at significantly higher pH values[11].

Although considerable progress has been made towards the elucidation of the regulatory aspects of solvent production, the details are still poorly understood. However, it becomes clear that the regulation takes place on several levels involving multiple regulatory factors, such as the regulatory proteins described above (Fig. 1.6).

The hierarchical regulation is underlined by potential cross-regulatory mechanisms between the different regulators – indicated by the presence of binding sites in the promoter regions of the respective other regulatory genes. For instance, 0A boxes are located in front of the *codY* gene as well as the *adcR/S* operon. Additionally, at least one potential CodY binding site in the promoter region of the *adcR/S* operon has been identified. The complexity of the regulation of solventogenesis is corroborated by the fact that a gene or operon is often regulated by a multitude of regulators, e.g. the *adc* gene or the *sol* operon (indicated by the different binding sequences in Fig. 1.6). Moreover, the presence of a variable number of binding sites of a regulator in the promoter region of a specific gene, e.g. two Spo0A binding sites in the promoter region of the *bdhA* gene versus one in front of *bdhB* (Fig. 1.6), could indicate a more or less significant influence of this protein in the regulation and increases the intricacy of the underlying regulatory mechanisms even further.

The development of the ClosTron gene knock-out system by Heap *et al.* in 2007 constitutes a major accomplishment in clostridial research[25]. Thenceforward, the generation of stable single and multiple regulatory mutants is possible and should lead to further important conclusions concerning this complex regulatory network.

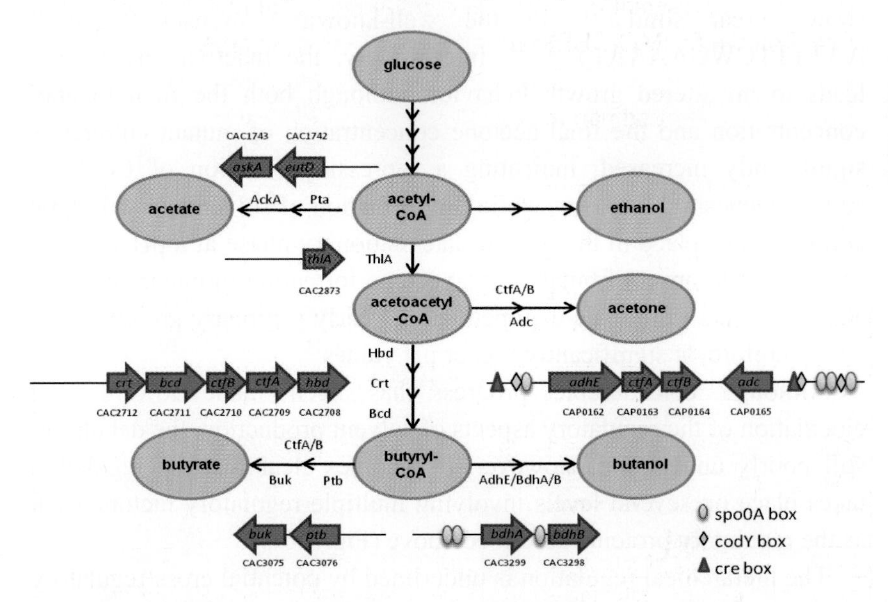

Fig. 1.6: Schematic overview of the fermentative metabolism of *C. acetobutylicum* as well as the genomic organization of involved genes and potential and experimentally confirmed binding sites of Spo0A, CodY and CcpA. Note: genes are not to scale!

While the construction of appropriate mutants to allow observation of cross-regulation mechanisms is underway (Chapter 2), modeling approaches concentrate on the development of hypotheses for regulation and the construction of modeling modules. Using the former approach, pH change as a triggering event for initiation of solventogenesis is postulated. Such a model has been developed and verified, however, without taking into account the detailed molecular interactions during regulation (Chapter 4). The latter approach aims at development of models for regulatory network modules which are possibly species independent. Characterization of these modules, creating a modeling library of various interactions will allow quick allocation of regulatory model structure once the underlying molecular mechanism is clarified (Chapter 6).

Supplementary material

A SBML representation of the genome-wide metabolic network can be found on the SysMO-SEEK website (seek.sysmo-de.org).

List of variables and parameters

	Description	**Units**
V_R	Reactor working volume (solid-liquid two-phase system)	m^3
V_L	Liquid volume in reactor, extracellular	m^3_{liquid}
V_C	Total intracellular liquid volume of cells in the reactor	m^3_{cells}
c_A	Reactor volume-based concentration of component A	mol A $(m^3)^{-1}$
$c_{A,ex}$	Extracellular concentration of component A	mol A $(m^3_{liquid})^{-1}$
$c_{A,in}$	Intracellular concentration of component A	mol A $(m^3_{cells})^{-1}$
c_X	Biomass concentration, dry weight per total reactor volume	kg_{DW} A $(m^3)^{-1}$
k_X	Density of cells as dry weight per cell volume	kg_{DW} A $(m^3_{cells})^{-1}$
r_A	Reaction rate for A related to reactor working volume V_R	mol A $(m^3As)^{-1}$
$r_{A,in}$	Intracellular reaction rate for component A	mol A $(m^3_{cells}As)^{-1}$
n_A	Transport flux of component A between intracellular and extracellular volume, related to V_R	mol A $(m^3As)^{-1}$
μ	Specific growth rate of biomass	s^{-1}
T	Time	s
$k_{i,j}$	Reaction rate constant for reaction i -> j	s^{-1}
$c_{Enzyme,i,j}$	Enzyme concentration catalyzing reaction i -> j	mol A $(m^3)^{-1}$
c_i, c_j, c_k	Concentrations of metabolites i, j and k	mol A $(m^3)^{-1}$
$K_{M,i,j}$	Michaelis–Menten constant for reaction i -> j	mol A $(m^3)^{-1}$
$K_{I,i,j,k}$	Inhibition constant for reaction i -> j by inhibitor k	mol A $(m^3)^{-1}$
$r_{i,j}$	Reaction rate for enzyme reaction i -> j	mol A $(m^3As)^{-1}$

References

1. Senger R. S. and Papoutsakis E. T., *Biotechnol. Bioeng.* 101 (2008a).
2. Senger R. S. and Papoutsakis E. T., *Biotechnol. Bioeng.* 101 (2008b).
3. Lee J. *et al.*, Appl. Microbiol. Biotechnol. 80 (2008).
4. Gheshlaghi R. *et al.*, *Biotechnol. Adv.* 27 (2009).
5. Ravagnani A. *et al.*, *Mol. Microbiol.* 37 (2000).
6. Harris L. M. *et al.*, *J. Bacteriol.* 184 (2002).
7. Bahl H. *et al.*, Appl. Environ. Microbiol. 52 (1986).
8. Girbal L. and Soucaille P., *Trends Biotechnol.* 16 (1998).
9. Fontaine L. *et al.*, *J. Bacteriol.* 184 (2002).
10. Schiel B. *et al.*, *Biospektrum special issue* 2003 KB003 (2003).
11. Standfest T. *et al.*, *Biospektrum special issue* 2009 PS 21 (2009).
12. Schiel B., Ph.D. Thesis, University of Ulm, Germany (2006).
13. Grimmler C. *et al.*, *J. Mol. Microbiol. Biotechnol.* 20 (2011).
14. Brehm S., Diplom Thesis, University of Ulm, Germany (2007).
15. Rodionov D. A. *et al.*, *FEMS Microbiol. Lett.* 205 (2001).
16. Tangney M. *et al.*, *J. Mol. Microbiol. Biotechnol.* 6 (2003).
17. Yu Y. *et al.*, *Appl. Environ. Microbiol.* 73 (2007).
18. Ren C. *et al.*, *Metab. Eng.* 12 (2010).
19. Grimmler C. *et al.*, *J. Biotechnol.* 150 (2010).
20. Nold N., Ph.D. Thesis, University of Ulm, Germany (2008).
21. Guédon E. *et al.*, *Microbiol.* 151 (2005).
22. den Hengst C. D. *et al.*, *J. Biol. Chem.* 280 (2005).
23. Belitsky B. R. and Sonenshein A. L, *J. Bacteriol.* 190 (2007).
24. Heap J. T. *et al.*, J. Microbiol. Methods 70 (2007).

CHAPTER 2

CLOSTRIDIAL GENE TOOLS

S. T. CARTMAN, J. T. HEAP, S. KUEHNE, C. M. COOKSLEY, M. EHSAAN, K. WINZER, N. P. MINTON

Centre for Biomolecular Sciences, BBSRC Sustainable Bioenergy Centre Digestive Diseases Centre NIHR Biomedical Research Unit, School of Molecular Medical Sciences University of Nottingham, University Park, Nottingham, NG7 2RD, UK

In recent years the genus *Clostridium* has assumed greater prominence in terms of both the diseases individual species cause and as a consequence of the renewed importance of certain strains in the production of chemical commodities from renewal biomass, and in particular biofuels. This has precipitated the determination of a plethora of genome sequences, and, more importantly, an acceleration in efforts directed at the development of genetic systems to facilitate the exploitation of the data being generated. As a consequence, new methods have been formulated with which directed mutants can be made to assist reverse genetic studies, most notably insertional systems based on retargeting group II introns, e.g. the ClosTron. In parallel to this, inroads have been made towards the development of efficient, random mutagens for forward genetic approaches, in the form of exemplification of transposome technology and a *mariner*-based transposon in *Clostridium perfringens* and *Clostridium difficile,* respectively. Still to be implemented are rapid and reliable methods for creating in-frame deletions as well as more effective methods for delivering large segments of DNA into the genome. The formulation of such methods has already reached an advanced stage of development in our laboratory. In the meantime, there is still some way to go before clostridial researchers can boast the degree of sophistication available to those researchers studying *Bacillus.*

2.1 Introduction

The large, diverse genus *Clostridium* consists of Gram-positive, anaerobic, endospore-forming, rod-shaped bacteria. Among these are historically infamous pathogens such as *Clostridium botulinum* and *Clostridium tetani*, as well as benign strains that have been used in industrial fermentations, most notably *Clostridium acetobutylicum* and *Clostridium beijerinckii*. The genus has been relatively little-studied during the molecular biology era, but recent years have seen a surge of interest as clostridial strains both old and new have become relevant to pressing social issues: high oil prices and the need to move towards a sustainable energy economy has revived interest in the use of clostridial fermentations for biofuel production; *Clostridium difficile* has gone from total obscurity to 'superbug' scourge of healthcare facilities worldwide; and necrotic enteritis in poultry, caused by *Clostridium perfringens*, has become a major problem since regulation has reduced antibiotic supplementation of animal feed. With the increasingly high profile of the genus, the need for mature genetic methodologies is clear.

From the 1970s onwards, there have been numerous reports of the transfer of specifically constructed recombinant plasmids into a range of *Clostridium* species[1-6], by either transformation or conjugative transfer from heterologous donors. As with many other prokaryotic systems, the vectors deployed are invariably shuttle plasmids that incorporate a selectable marker and both a Gram-negative (most usually, but not exclusively, the *E. coli* ColE1 replicon) and a 'clostridial' Gram-positive replicon. A number of specialist vectors (e.g. promoter probe vectors or expression plasmids) have also been constructed and tested. Successes in the design and transfer of basic plasmid systems have until recently not been mirrored by the deployment of systems for the reproducible insertion of DNA into the bacterial genome. Such systems are essential for ascribing function to those genes identified by genome sequencing, as well as for the advanced engineering of desired enhancements to commercially important strains.

2.2 Clostridial gene transfer

A number of different *E. coli*/*Clostridium* shuttle vectors were developed during the 1980s and 1990s, and a number of reviews describing them have appeared[3,5]. Transfer was elicited by either transformation or by conjugative transfer from a heterologous donor. In the early days, transformation was in some cases achieved by polyethylene glycol (PEG)-induced uptake of DNA by cells stripped of their cell walls and the subsequent regeneration of vegetative cells from the transformed protoplasts using specially formulated, species-specific, regeneration media[4]. However, with the advent of electroporation, such methods fell into disuse, and electroporation now represents the most commonly used transformation procedure[5]. In particularly recalcitrant species, such as *Clostridium thermocellum*, this has necessitated the fabrication of purpose-built electroporation apparatus[7].

In many *Clostridium* species, however, conjugative plasmid transfer represents either the only (e.g. *C. difficile*) or the most efficient (e.g. certain species of *C. botulinum*) means of achieving plasmid transfer. The process requires close cell-to-cell contact and involves both a *cis*-acting nick site (*oriT*, origin of transfer), a number of *trans*-acting functions that are necessary for mating pair formation as well as DNA processing and transfer of the conjugative plasmid to the recipient cell[8]. Whilst in the early days donor species such as *Bacillus subtilis, Lactococcus lactis* or *Enterococcus faecalis*[9,10] have been used as the conjugative donor, the preferred host is now invariably *E. coli*. In this instance, the shuttle vector is most commonly mobilised through the provision of the *oriT* region of an IncP plasmid such as RP4 or RK2[11]. This region, in conjunction with several additional trans-acting functions (Tra functions), is absolutely required for the conjugative transfer of the plasmid. The Tra functions are provided by the *E. coli* (Tra^+) donor, and may either be plasmid-encoded (e.g. carried by a IncP-type helper plasmid such as R702), or integrated into the chromosome, as is the case with *E. coli* strain SM10. In our hands at least, the former strategy has invariably proven to be most effective, and the *E. coli* donor strain CA434 (*E. coli* HB101 carrying the plasmid R702) developed by the Young laboratory[11] represents the most efficient donor in most clostridial conjugations.

In most instances, the simplicity of conjugative protocols and the high frequency obtained, coupled with the lack of a requirement for a specialised electroporation apparatus, make conjugation the preferred method. However, despite extensive testing, certain clostridial species/strains appear to be unable to act as conjugative recipients with

E. coli donors. The most notable example is *C. acetobutylicum* ATCC 824, where such transfer has never been demonstrated, despite the observed ability of this strain to act as a recipient in the transfer of conjugative transposons from *Bacillus subtilis, Lactococcus lactis* or *Enterococcus faecalis* donors[9,10]. The reason why is unclear, but may suggest that the required cell-to-cell contact between these two species does not occur.

Whilst it is possible that the absence of appropriate receptors on the *C. acetobutylicum* cell surface precludes the formation of the required 'conjugative bridges' with an *E. coli* donor, in other instances failure to achieve transfer has been due to a restriction barrier. In these instances, the presence of a specific restriction activity in the recipient and the absence of appropriate donor-mediated methylation of the incoming plasmid DNA, prevent the successful establishment of the transferred plasmi'd. A barrier of this type has traditionally been overcome by experimentally determining the specificity of the restriction activity present in the recipient, and then endowing the donor with an appropriate methylase gene which, when expressed, protects the DNA from subsequent cleavage by methylation of the target recognition restriction site. Such a strategy was successfully employed to achieve DNA transfer to *C. acetobutylicum* ATCC 824[12], *C. cellulolyticum* ATCC 35319[13], *C. botulinum* ATCC 25765[14] and *C. difficile* CD3 and CD6[15]. With the advent of the genome era, the task has been considerably simplified as methylase genes in the recipient may be rapidly identified *in silico*, allowing them to be cloned and expressed in the intended donor cell.

2.3 Clostridial vector systems

As with many other prokaryotic systems, the vectors deployed within the genus are invariably shuttle plasmids that incorporate a selectable marker and both a Gram-negative and a 'clostridial' Gram-positive replicon. Although a reasonably wide number of Gram-positive plasmid replicons have been investigated and shown to function in the various clostridial species tested, only a handful have found extensive favour. In some instances, the replicon used has been derived from a clostridial source. In other cases, the replicon has been derived from a non-clostridial,

Gram-positive species. Of particular note are the replicons of the *C. perfringens* plasmid pIP404 (widely used in *C. perfringens*), the *C. difficile* plasmid pCD6 (widely used in *C. difficile*), the *C. butyricum* plasmid pCB102 (widely used in *C. beijerinckii*), the *Enterococcus faecalis* plasmid pAMβ1 (widely used in *C. beijerinckii* and *C. cellulolyticum*) and the *Bacillus subtilis* plasmid pIM13 (widely used in *C. acetobutylicum*)[3,5]. Interestingly, the latter plasmid has also been found to be applicable to *C. thermocellum*, indicative of an intrinsic ability to replicate at high temperature[7].

Other considerations in the design of clostridial vectors have been the provision of markers, through which the presence of the plasmid can be initially selected upon transfer and subsequently maintained[16]. Here there is a dearth of available elements for this purpose. Genes specifying resistance to erythromycin and/or lincomycin (*erm*) have proven to be of greatest utility, closely followed by *cat* genes specifying resistance to chloramphenicol/thiamphenicol. After this, choices are very limited. In certain species *aad* genes (resistant to spectinomycin) or *tet* genes (resistant to tetracycline) can be used, but they are not applicable in all cases, and where they can be employed are often disadvantaged by problems of background growth on selective plates. There is a clear need for the identification of additional markers. Maintenance through selection is an important consideration, as all clostridial vectors constructed to date exhibit varying degrees of segregational instability, and will therefore be lost from the population in the absence of selective pressure.

The shuttle plasmids generally available do not share a common structure, reflecting their construction in different laboratories, and were largely assembled with little emphasis on a design that can easily be modified. Consequently, altering these plasmids tends to require *ad hoc* cloning strategies which may prove difficult and time consuming. Furthermore, it is not possible to directly compare the differing functional properties of these plasmids. This can cause problems, especially for research groups who work on several different *Clostridium* species. Recently, therefore, we took a more systematic approach to vector design, and created the pMTL80000 modular shuttle plasmids[16]. Essentially, we defined a standard arrangement in which every plasmid

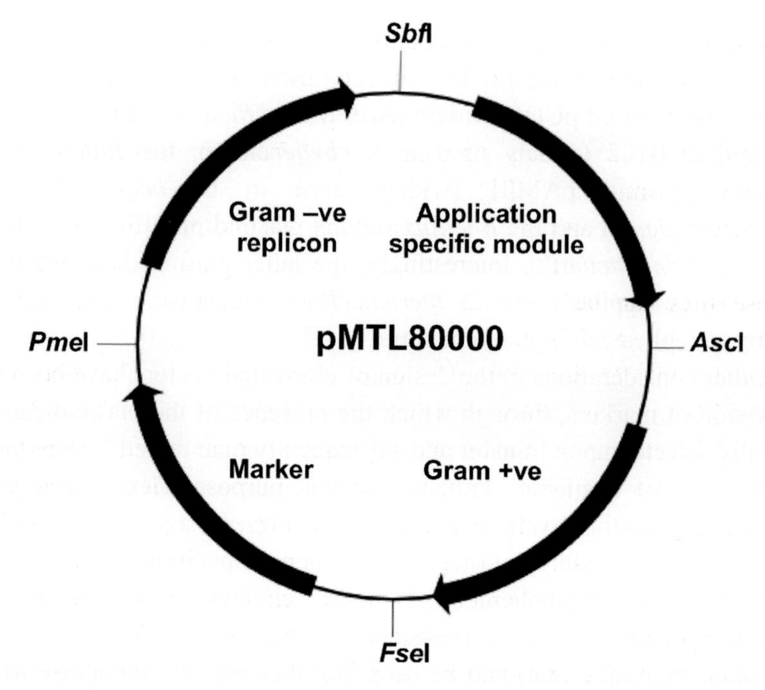

Fig. 2.1: Schematic view of the pMTL80000 modular vector series. The plasmids have a standard arrangement in which exactly one module of each of four types is always arranged in the same order, and always bounded by the same four rare (8 bp) type II restriction enzyme recognition sites indicated[16]. The modules are numbered and listed in Table 2.1.

contains exactly one module of each of four types, always arranged in the same order, and always bounded by the same four rare (8 bp) type II restriction enzyme recognition sites (*Asc*I, *Fse*I, *Pme*I and *Sbf*I). The standard arrangement is shown in Fig. 2.1. We also constructed a total of 18 numbered modules of these four types, flanked by the appropriate restriction sites. This system allows the combinatorial construction of shuttle plasmids from modules in the standard format. It also provides for the quick and easy modification of existing pMTL80000-based plasmids.

When constructing plasmids using the existing modules no real design process is required — the researcher need only choose the desired combination of modules from the available selection (see http://www.clostron.com/pMTL80000.php). Due to the standard construction, the sequence of pMTL80000 plasmids is precisely defined by the choice of modules, and so can easily be determined automatically. Plasmid names are also determined by the combination of modules: 'pMTL8' followed by the numbers of the combination of modules used, in the correct order. Consequently, the short plasmid names each conveniently imply a precise sequence (Table 2.1).

Annotated sequences of any particular pMTL80000 plasmid can be downloaded from http://www.clostron.com in Genbank (.gb) format. These files can be interrogated using appropriate molecular biology/bioinformatics tools such as GENtle, ApE and Vector NTI. In order to distribute the 18 modules in a convenient way, five plasmids were constructed (pMTL80110, pMTL82254, pMTL83353, pMTL84422 and pMTL85141) which between them contain all 18 modules. Through supply of the 'core set' of five plasmids as a kit (free-of-charge to academics), any of the other 395 possible combinations of modules can be easily constructed in one or a few cloning step(s) using *Asc*I, *Fse*I, *Pme*I and *Sbf*I (and optionally in some cases *Apa*I, where provision of *oriT* is required).

Once we had established the pMTL80000 system, we were able to generate directly-comparable data on the practically-relevant functional properties of the Gram-positive replicons and antibiotic markers commonly used in *Clostridium* (Table 2.2). Using these results and the pMTL80000 system, it is now possible to quickly construct shuttle plasmids for particular hosts and applications using rationally-chosen combinations of modules[16]. The system has been extensively tested in our laboratory, where it is utilised in all ongoing recombinant work. We propose that adoption of this modular system as a standard would be of substantial benefit to the *Clostridium* research community, whom we invite to use and contribute to the system.

2.4 Tools for forward genetics studies

Pivotal to the molecular dissection of physiological traits is the ability to make mutations. In the absence of clues as to the nature of the genes and

Table 2.1: The modular components of the pMTL80000 vector series.

Modular Plasmid CORE Set[a]	**Numeric Module Assignment**			
	Gram +ve Replicon	**Marker**	**Gram –ve Replicon**	**Application Specifc**
pMTL80110	0. spacer	1. *catP*	1. p15a	0. spacer
pMTL82254	2. pBP1	2. *ermB*	5. ColE1 + *oriT*	4. *catP* reporter
pMTL83353	3. pCB102	3. *aad9*	5. ColE1 + *oriT*	3. P_{fdx} + MCS
pMTL84422	4. pCD6	4. *tetA*	2. p15a + *oriT*	2. P_{thl} + MCS
pMTL85141	5. pIM13	1. *catP*	4. ColE1	1. MCS

[a] The indicated plasmids are the 'core' set, the distribution of which provide researchers with all 18 existing modules, providing the facility to construct all possible variants. The plasmids are named according to the modules present. Thus, plasmid pMTL83353, carries the Gram-positive replicon of the plasmid pCB102, the *aad9* gene, the ColE1 Gram-negative replicon and the *oriT* region of plasmid RK2, and a region encompassing multiple cloning sites (MCS). Other components are described at http://www.clostron.com[16].

their products involved in any particular process, mutants of a desired phenotype may be generated randomly and thence the genotype responsible determined. This process of 'forward genetics' is reliant on the availability of effective random mutagens, most commonly transposons. Until very recently, the only such elements available were conjugative transposons.

2.4.1 Conjugative transposons

Conjugative transposons are mobile elements that possess the genetic machinery to facilitate their own transfer between bacterial cells[17]. They exist in the main as chromosomally integrated DNA elements that excise themselves from the genome to form a covalently closed circular intermediate. This intermediate can either reintegrate in the same cell

Table 2.2: Properties of a selection of different modular vectors.

Plasmid	Source of Replicon	***Clostridium acetobutylicum*** **ATCC 824**		***Clostridium botulinum*** **ATCC 3502**		***Clostridium difficile*** **CD 630**	
		Transfer Frequency [a]	Segregational Stability [b]	Transfer Frequency [a]	Segregational Stability [b]	Transfer Frequency [a]	Segregational Stability [b]
pMTL82151	pBP1	1.38×10^{2}	99.4±0.9	1.01×10^{-3}	99.9±0.3	3.36×10^{-6}	87.3±1.3
pMTL83151	pCB102	2.45×10^{2}	76.5±6.0	2.90×10^{-4}	99.4±0.3	2.23×10^{-6}	76.2±0.5
pMTL84151	pCD6	8.47×10^{1}	82.4±9.6	5.71×10^{-6}	81.6±3.7	7.00×10^{-6}	77.4±2.1
pMTL85151	pIM13	2.92×10^{2}	81.6±8.6	7.80×10^{-6}	89.6±0.7	4.18×10^{-7}	69.0±1.1

[a] - For *C.acetobutylicum* ATCC 824, the values shown are transformant colonies per µg plasmid DNA. For *C.botulinum* ATCC 3502 and *C.difficile* 630, the values shown are transconjugant colonies per *E.coli* donor Colony Forming Units (CFU). [16]

[b] - Segregational stability per generation is equal to $\sqrt[n]{R}$ where R is the fraction of cells containing the plasmid at the latest timepoint at which it could be determined, and n is the number of generations of growth without selection by this timepoint. Given the 1% inocula used, and assuming the cultures reach maximum cell density in 12 h, we took the number of generations per 12 h to be 6.64 (because $1 \times 2^{6.64} = 100$). For exact methodology, see [16].

(intracellular transposition) or transfer by conjugation to a recipient and integrate into the recipient's genome (intercellular transposition). They are by necessity large elements (16 kb to 75 kb and larger) and in certain clostridia play an intimate role in the spread of antibiotic resistance, most notably for *C. difficile*[18-21]. One of the elements involved, Tn*5397*, has been well characterised[22,23] and has been found to share substantive similarity to Tn*916*, essentially differing only in their respective integration and excision systems. As Tn*916* has found widespread use in other bacteria as a mutagenic tool, its use in forward genetic studies has been extensively investigated in both pathogenic and non-pathogenic *Clostridium* species.

In pathogenic clostridia, Tn*916*, has been used to obtain a variety of mutants, including auxotrophic mutants of *C. botulinum*[24] and *C. perfringens*[25], as well as pleiotropic mutants of the latter affected in toxin production[25,26]. In the case of *C. perfringens,* subsequent analysis revealed that the effects on toxin production were a consequence of insertion into a gene encoding the sensor kinase *virS*, part of a two-component system (VirRS) now known to regulate the expression of virulence factor genes[27]. In both pathogens, some defects in toxin production were due to large, apparently Tn*916*-mediated, deletion events[25,28]. Initial results with Tn*916* in the non-toxinogenic *C. difficile* strain, CD37, indicated that the element inserted at a preferred site, or 'hot-spot'[29,30], thereby precluding its use as a random mutagen. However, it was subsequently demonstrated that Tn*916,* and its derivative Tn*916*ΔE, integrates at random in other *C. difficile* strains[31,32], including the genome sequence strain, CD630[23].

The use of conjugative transposons in the non-pathogenic bacteria has principally focused on their use in obtaining mutants defective in solvent production and/or sporulation. In *Clostridium acetobutylicum* and *C. saccharobutylicum* P262 (formerly *C. acetobutylicum*) the transposon employed has been Tn*916*, and the donor *E. faecalis*[33-35]. However, the mutants obtained remained largely uncharacterised. These were a Tn*916* insertion into *thrA* (encoding $tRNA^{Thr}$), suggesting that the use of the codon ACG has a regulatory role in solvent production[36], and the identification of a *sum* gene, apparently essential for clostridial differentiation and sporulation in *C. saccharobutylicum* P262[37].

In contrast to the other solventogenic species, *C. beijerinckii* NCIMB 8052 (formerly *C. acetobutylicum*) has a hot-spot for Tn*916*[10], necessitating the use of Tn*1545* for mutant generation[38,39]. These included the isolation of a mutant deficient in 'degeneration' (the loss in prolonged culture of the ability to produce solvents) and mutants exhibiting increased butanol tolerance[40]. The former was subsequently found to be merely a consequence of a reduction in growth rate due to a partial inactivation of the gene (*fms*) encoding peptide deformylase[41], while the latter was hypothesised to be caused by a reduction in the activity of *gldA* (glycerol dehydrogenase) due to production of anti-sense RNA from a Tn*1545* located promoter. Why a reduction in the encoded glycerol dehydrogenase activity should result in an increase in butanol tolerance is unclear.

In general, however, one of the major drawbacks encountered with conjugative transposons as a mutagenic tool is their predilection to insert in multiple copies. In the case of *C. acetobutylicum* in the studies of Bertram and Dürre[33] and Mattsson and Rogers[34], 50% or more of the transconjugants isolated had more than one transposon insertion. In *C. botulinum*, of the 13 transconjugants analysed, only six were found to carry a single copy of the transposon, with the other seven carrying either two or three copies[24], while in *C. difficile*, six out of 10 transconjugants analysed contained more than one copy of the element[32]. Only one of the *C. perfringens* mutants isolated by Awad and Rood[42] contained a single insertion of Tn*916*, multiple Tn*916* insertions accounted for between 65–75% of the *C. perfringens* mutants isolated by Kaufmann and coworkers[25], and only 12 of 42 independently isolated *C. perfringens* mutants had single copy insertions of Tn*916*[43].

2.4.2 Non-conjugative transposons

The use of conjugative transposons is disadvantaged by poor efficiency, their large size and, most unfavourably, their predilection to insert in multiple copies. A number of non-conjugative transposon mutagenesis systems have been described for Gram-positive bacteria[44–47], but until recently no systems existed for use in *Clostridium* species. The situation has now changed with the development of two independent transposome systems for *C. perfringens*, and the adaptation of a *mariner*-based transposon for *C. difficile.*

2.4.2.1 Transposome mutagenesis systems

The transposome approach involves the *in vitro* assembly of a complex between the transposon DNA and the transposase enzyme, the transpososome, followed by delivery of the transpososome into the recipient cells. Once inside a cell, the transpososome becomes activated in the presence of divalent cations, resulting in genomic integration of the delivered transposon. The first system to be implemented in *C. perfringens* was based on a bacteriophage Mu[48], where 239 and 134 transposon insertions per ug of DNA were obtained in a laboratory strain (JIR325, a derivative of strain 13) and a field isolate (strain 56) of *C. perfringens,* respectively. However, whilst sequencing revealed that protein-encoding genes comprised 44.5% of the integration sites and 12.5% in intergenic regions, a high percent (43%) carried transposon insertions into one of the rRNA genes. This latter preference for rRNA genes somewhat limits the utility of the system. Subsequent to this study, a similar approach was taken using a transposome based on an EZ-Tn5 random mutagenesis system (Epicentre®). All of the EZ-Tn5 mutants obtained were shown to contain only a single transposon insertion[49]. Moreover, the transposon mutants were produced at a 46-fold higher frequency than that obtained using the Mu system, and led to the identification of a homologue of the staphylococcal *agr* system shown to regulate early production of both alpha-toxin and perfringolysin O (PFO) by strain 13[49].

2.4.2.2 A mariner*-based mutagenesis system*

Whilst both of the above transposome systems represented a considerable improvement on the use of conjugative transposons, in terms of the absence of multiple insertions, they are absolutely reliant on attaining high frequencies of transformation in the target organism. Whilst the transformation frequencies obtained in *C. perfringens*, and in particular strain 13, are relatively high, such frequencies are relatively rare in other *Clostridium* species. Moreover, in many clostridia, such as *C. difficile*, transformation has never been demonstrated, and DNA transfer by conjugation represents the only means of introducing plasmids. In order to develop a system more applicable to species such as *C. difficile*, we turned our attention to the development of a *mariner*-based transposon[50].

The *mariner* transposable element *Himar1* has been shown to insert randomly into the genomes of many bacterial species[45–47,51,52], where the cognate *Himar1* transposase is the only factor required for transposition. Transposons of this type are defined by inverted terminal repeats (ITRs) at either end and insert into a 'TA' target site through a 'cut-and-paste' mechanism[53]. To adapt this system to *C. diffficile*, the plasmid pMTL-SC1 (Fig. 2.2) was constructed in which the transposase gene was placed under the control of the *tcdB* promoter. This strategy had a number of advantages. In the first instance, transcription from *tcdB* is reliant on a specialised class of sigma factor, unique to the toxinogenic clostridial species, *C. botulinum*, *C. difficile* and *C. perfringens*[54]. Accordingly, expression of the transposase in the *E. coli* donor, which lacks this sigma factor, is prevented, thereby ensuring stability of the element in *E. coli* both during the construction of the plasmid and prior to its transfer to the clostridial recipient. Secondly, *tcdB* is catabolite repressed[55], providing the facility to limit, in the early stages of transfer, expression of the transposase through supplementation of the media with glucose. This facility improves the properties of the library — preventing a single early transposition event leading to a clonal population with the same insertion.

Whilst traditional transposon mutagenesis procedures rely on the transfer of suicide vectors, such a strategy is not applicable to *C. difficile*, due to the low frequencies of transfer. To circumvent this, we made use of the principle of 'pseudo-suicide', in which a vector is chosen that replicates extremely poorly. Such plasmids can become established and maintained in the cell population through imposition of antibiotic selection, but they are rapidly lost in the absence of antibiotics. The identification of the optimum replicon for this purpose was established prior to the construction of pMTL-SC1 using the pMTL80000 modular vector series[16]. Accordingly, each replicon was tested in an identical

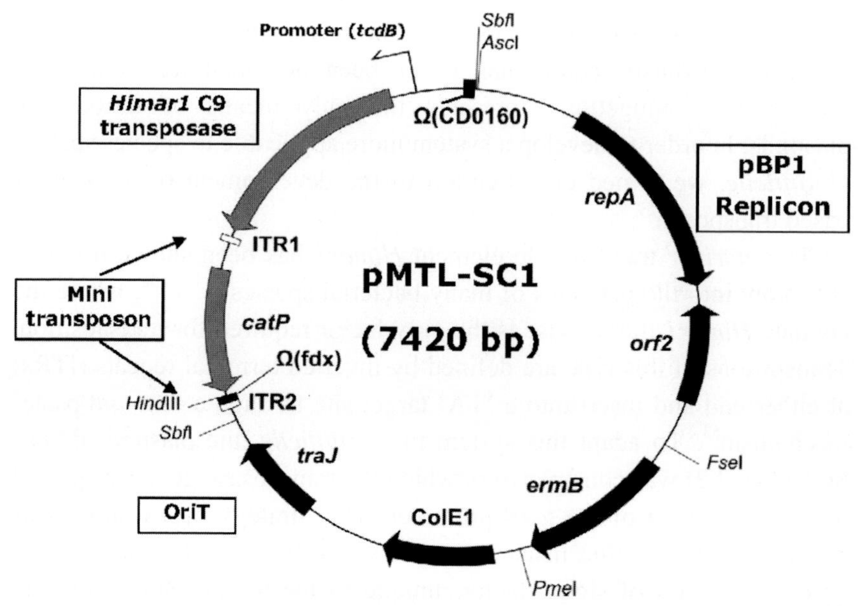

Fig. 2.2: A mariner-based transposon delivery system exemplified in *C. difficile*[50]. Expression of the transposase gene *Himar1 C9* is under the control of *C. difficile* toxin B gene (*tcdB*) promoter. The plasmid backbone consisted of the pBP1 replicon of *C. botulinum* (*repA* and orf2), the enterococcal *ermB* gene (conferring resistance to rythromycin), the Gram-negative ColE1 replicon and the conjugal transfer function *traJ* of the *oriT* region of plasmid RK2. The transcriptional terminators (Ω) are identical in sequence to those found immediately downstream of the ferredoxin gene of *C. pasteurianum* and the CD0164 open reading frame of *C. difficile* 630. This vector conforms to the pMTL80000 modular system for *Clostridium* shuttle plasmids[16].

shuttle vector context (incorporating *catP*, ColE1, *oriT* and a *lacZα* gene) in *C. difficile* R20291[50]. Not unexpectedly, the pCD6-based plasmid appeared most effective, forming visible colonies after just 24 h on media supplemented with thiamphenicol. In contrast, cells harbouring either the pBP1 or the pCB102-based plasmid took 48–72 h to form colonies. We selected the pBP1 replicon as our pseudo-suicide plasmid replicon because, although slightly more stable (1.3% per generation) than pCB102, it could be conjugated into R20291 at a frequency almost eight-fold greater.

For library construction, the *mariner* plasmid pMTL-SC1 was transferred into *C. difficile* R20291 by conjugation. Transconjugants were initially selected on BHIS supplemented with glucose and the antibiotics cycloserine, cefoxitin and lincomycin (selecting for both *C. difficile* and pMTL-SC1, and counter-selecting against *E. coli*), then picked and restreaked onto TY medium supplemented with the same antibiotics, but lacking glucose in order to enhance expression of *Himar1* C9 transposase from the *tcdB* promoter. After 72 h, all growth was harvested into PBS, serial dilutions were made and plated onto BHIS supplemented with cycloserine, cefoxitin and thiamphenicol, to select for the transposon-based *catP* marker. Transposon insertions arose at a frequency of $4.5(\pm 0.4) \times 10^{-4}$ (calculated as the ratio of Tm^R cfu to total cfu). Individual colonies, visible after 12–16 h, were picked and restreaked onto the same medium twice for further analysis and/or phenotypic screening. Nucleotide sequencing of the site of insertion of 60 random mutants demonstrated that transposon insertions occurred at a 'TA' di-nucleotide target site, and were distributed throughout the genome of *C. difficile* R20291, with no evidence for a preferred target site. Overall, there were 28 insertions in the plus (+) strand and 32 in the minus (–) strand. Moreover, 75% (45 of the 60 insertions sequenced) were located within protein coding sequences. As 81% of the genome is protein coding, this is within the range that would be expected for a random mutagen. The utility of the system for forward genetic studies was demonstrated through the isolation of mutants defective in germination (insertion in a putative germination-specific protease gene, *cspBA*) and uracil metabolism (insertion in the gene encoding the aspartate carbomoyltransferase catalytic chain, *pyrB*). Collectively, these data establish that our *mariner*-based transposon system is an effective tool for generating libraries of random mutants in *C. difficile*, and suggest that with appropriate adaptation, the system should prove to be applicable to other clostridial species.

2.5 Recombination-based tools for reverse genetic studies

The directed inactivation of genes is pivotal to functional genomic approaches, and is most often accomplished through procedures reliant on homologous recombination. In essence, an inactivated copy of the gene being targeted is introduced into the cell on an extra-chromosomal element, where recombination between the inactivated form and the wild-type copy of the chromosomal gene occurs. When recombination occurs across two regions of homology (double cross-over), the intervening segment of DNA is inserted into the genome, replacing the equivalent wild-type region by reciprocal exchange. If a single recombination event takes place (single cross-over), the entire element integrates into the genome by a Campbell-like mechanism. If the net result of either process is to be mutagenic, then different types of 'knock-out' cassette need to be constructed. For double cross-over, the fragment that replaces the wild-type region needs to be inactivated through either insertion or deletion of DNA. For single cross-over insertions, the introduced DNA needs to encompass a central portion of the gene, such that upon integration, the resultant two duplicated copies of the targeted gene are non-functional due to the absence of either the 5'-end or the 3'-end of the gene. An inherent consequence of the ensuing duplication is that the mutants generated are unstable, as recombination between the two copies of the gene will result in excision of the plasmid.

To allow selection of the relatively rare recombination events, it is most common to use plasmids which are incapable of replication (so-called suicide vectors) and which carry a selectable antibiotic resistance gene. In the presence of an antibiotic, clonal cells may only arise if the marker gene integrates. In single cross-over approaches, where the entire plasmid integrates, the marker gene can reside on the vector backbone. In double cross-over strategies, the marker gene is most conveniently placed centrally within the region to be exchanged, where it is responsible for inactivation of the gene. In this case, a second marker (e.g. another antibiotic resistance gene) needs to be placed elsewhere on the vector so that loss of the vehicle can be determined. Alternatively, to create non-polar mutants, inactivation of the gene is accomplished through a simple in-frame deletion rather than inserting an antibiotic resistance gene, and a counter-selectable marker is incorporated into the vector backbone in addition to the antibiotic resistance marker. In this configuration, the attainment of a double cross-over is achieved in two steps. In the first, integration of the plasmid via a Campbell-like event is positively selected by acquisition of antibiotic-resistance. Thereafter, excision of the plasmid by a second recombination event is selected by loss of the counter-selection marker, yielding wild-type revertants and the desired deletion mutants in similar proportions. One of the most widely used of such negative selection markers is the *Bacillus subtilis sacB* gene, which confers lethality on the host in the presence of exogenous sucrose[56].

2.5.1 Mutants in pathogenic clostridia

In terms of mutant generation by recombination-based methods, the greatest success has been achieved to date in *C. perfringens* strain 13; a consequence of its ability to be transformed at high efficiencies. This made possible the deployment of simple suicide vectors (replication deficient) carrying a copy of the gene to be inactivated into which was centrally inserted either a *tet*[57] or an *ermBP*[58] selectable marker. In both studies, the vectors were designed with the intention of achieving reciprocal exchange, by double cross-over. Double cross-over mutants were distinguishable from non-mutagenic single cross-over insertions by simple phenotypic plate tests which measured the presence or absence of both toxins, e.g. haemolysis on horse blood agar in the case of *virR*[57] and *pfoA*[58]. The frequency of integration was shown to be directly related to the extent of homology employed, although the efficiency of the method was quite low, e.g. the number of θ-toxin mutants isolated represented between 1.9% and 2.8% of the total erythromycin resistant colonies obtained in three of four experiments[58]. In general, single cross-over, non-mutagenic events were found to be in the majority, but that rarer double cross-over, mutant integrants could be detected through appropriate screening. A two-step approach has become the method of choice in most subsequent attempts to generate double cross-over mutants, in which one antibiotic gene is used to inactivate the gene (e.g. *catQ*), and a second (e.g. *ermBP*) is placed on the vector backbone. This allows the initial selection and isolation of rare single cross-over integrants on the basis of erythromycin (*ermBP*) resistance, and then the subsequent screening of such a clone for the presence of chloramphenicol resistant and erythromycin sensitive derivatives within the population in which the plasmid, along with *ermBP,* has excised to yield the desired reciprocal exchange. This type of strategy was used to inactivate the *colA* gene of strain 13[59] and *cpe* strains SM101 and FM4969[60], a *plc* mutant of the chicken pathogen strain EHE-NE18[61] and a *csa* mutant in *C. septicum*[62].

In other studies, suicide plasmids were designed and deployed specifically to generate single cross-over mutants, e.g. in *pfoA*[63], in *pfoR*[64], *luxS* and *virX*[65], double *pfoA*, *plc* mutants of strain 13[66] and *pilT* and *pilC* mutants[67]. A single cross-over mutant of the *gldA* gene of *C. difficile* strain CD37 was the first clostridial mutant to be reported in this species[68], and was generated using a suicide vector, pMTL30, previously successfully deployed in saccharolytic clostridial species[69]. Subsequently to this, efforts in *C. difficile* have focused on the use of a replication-defective plasmid based on the pIP404 replicon, shown to replicate extremely inefficiently in *C. difficile*[15]. Two genes were initially inactivated, *rgaR* (CD3255) and *rgbR* (CD1089), through single cross-over integration of the plasmid[70]. The same strategy was subsequently used to independently inactivate the *codY* gene and the two toxin genes (*tcdA* and *tcdB*) of strain 630[70,71]. However, such approaches are disadvantaged by the inherent instability of the mutants generated, as recombination events between the two copies of the target gene generated after Campbell-like integration of the plasmid invariably lead to excision of the plasmid. Thus, Southern blotting of the *C. difficile* *rgaR* mutant revealed the presence of 'looped out', independently replicating plasmids in some cells in the population[70], while in the study of Dineen and coworkers[71], the culture became dominated by wild-type cells in the absence of antibiotic selection. A more comprehensive list of the types of mutant generated in pathogenic clostridia appears in a recent review[72].

2.5.2 Solventogenic clostridia

In the solventogenic clostridia, mutant generation has largely been accomplished by single cross-over gene knockouts using suicide plasmids[72]. Such a strategy was first successfully reported in *C. beijerinckii* NCIMB 8052 (formerly *C. acetobutylicum* NCIMB 8052) through inactivation of *gutD* (glucitol dehydrogenase) and *spo0A* using a mobilisable (*oriT)* suicide plasmid, pMTL30, endowed with an *ermB* gene for selective purposes[69]. Thereafter, an identical strategy[73] was used to create mutants in the *scrB* (sucrose hydrolase) and *scrR* (transcript-tional regulator). In the biobutanol organism, *C. acetobutylicum*, four mutations (*butK*, CA_C3075; *pta*, CA_C1742; *aad,* CA_CP0162; and *solR*, CA_CP061) have been made by single cross-over integration of a replication-deficient plasmid introduced by electroporation[74-76]. In a later study, Harris and coworkers[77] adopted a similar approach to the double cross-over strategy of Sarker *et al.*[60] to inactivate *spo0A*. Based on the replicon of pIM13, their plasmid (pETSPO) carried a chloramphenicol resistance gene and a knock-out cassette in which *ermB* interrupted the *spo0A* gene. However, the desired double cross-over event was not obtained. Rather, recombination occurred between the two 10-nt homologous sequences (5′-ACGACCAAAA-3′) that were present in the 3′ end of the structural gene and upstream of *ermB,* leading to the loss of the 3-kb intervening sequence and concomitant inactivation of *spo0A*.

Mutations generated in solventogenic clostridia have, therefore, generally been accomplished by single cross-over knock-out. As a consequence, as seen in *C. difficile,* the mutants created are unstable, e.g. losses per 30 generations of between 1.8 to 3.0×10^{-3} for *buk* and *pta* in *C. acetobutylicum*[74] and between 0.37 to 1.3×10^{-3} for *C. beijerinckii*[69].

2.5.3 Negative selection markers

Whilst the generation of mutants using recombination-based approaches is becoming more commonplace, it is clear that progress would be hastened if negative selection markers were readily available, such as *sacB*[56]. The ability to counter-select considerably aids in the generation of the most effective mutations, in-frame deletions. In this respect some progress has been made by Philippe Soucaille's laboratory, through the adaptation of the *B. subtilis* method described by Fabret and coworkers[78] for use in *C. acetobutylicum* (P. Soucaille, personal communication). The method is reliant on the sensitivity of cells to 5-Fluorouracil, a consequence of its conversion to 5-fluoro-dUMP by the product of the *upp gene,* uracil-phosphoribosyl-transferase. Deletion of *upp* from the *B. subtilis* genome, enabled the use of *upp* as a negative selection marker (selecting for acquired resistance to 5-Fluorouracil upon plasmid excision and loss) when localised to the knock-out vector backbone[78]. An identical strategy has now been applied to *C. acetobutylicum* ATCC 824 (P. Soucaille, unpublished data). However, the system is somewhat disadvantaged, particularly in terms of virulence studies in pathogens, as strains are always phenotypically Upp-minus, in addition to the mutation being analysed. It is clear that for either alternative, heterologous negative selection markers are required, or a simple means is formulated such that the *upp* defect can be converted back to wild-type once the secondary mutation has been made.

2.6 Reverse genetic tools based on recombination-independent methods

Whilst it is possible to generate mutants in clostridial species by recombination approaches, the methods employed are not applicable to all strains and species, and, with the exception of *C. perfringens,* are generally relatively inefficient. Even in *C. perfringens,* success has been largely confined to the highly transformable strain 13. One solution is to utilise a process which is largely independent of host recombination factors, typified by group II introns.

2.6.1 Group II introns

Group II introns are catalytic RNAs that excise themselves from RNA transcripts via a lariat intermediate. Initially identified in organelles of plants and lower eukaryotes, the genomics era has revealed their presence in an increasing number of prokaryotes[79]. Bacterial group II introns differ significantly from their eukaryotic counterparts, as they tend to be exon-less (residing outside of structural genes) and are closely associated with mobile elements[79]. Mobile group II introns carry an ORF specifying an intron-encoded protein (IEP). The IEP functions in both RNA splicing and insertion through the reverse transcriptase activity of its N-terminal domain, an X domain associated with splicing, and a C-terminal endonuclease domain involved in mobility.

The group II intron from within the *ltrB* gene of conjugative plasmids found in *Lactococcus lactis* (the '*L1.LtrB* intron') is one of the best-studied bacterial examples. The target specificity of the intron is largely determined by base-pairing between sequences within the intron RNA and the target site DNA. Through an elegant series of experiments, the Alan Lambowitz laboratory were able to define the nature of the interactions that govern target recognition and went on to demonstrate that the target specificity of the intron could be altered through the introduction of specific changes[80]. This subsequently led to the formulation of an algorithm whereby the changes necessary to redirect the *Ll.LtrB*-derived introns to a gene of interest could be reliably predicted[81]. A further pivotal finding was the demonstration that *ltrB* need not be located within the intron, but could be provided *in cis* or *in trans.* This allows *ltrB* to be positioned on the backbone of the group II intron delivery plasmid. Loss of this plasmid from the cell following insertion of the group II intron into its target gene ensures the insertion event is mutational by preventing LtrB-mediated splicing of the inserted intron sequences. Further mobility of the intron is also prevented.

The retargeted introns that have resulted from these studies are called 'Targetrons'. Some are distributed by Sigma-Aldrich, who also sell a kit which provides the template necessary to generate a small fragment of DNA (ca. 350 bp) needed to alter the specificity of the Targetron by PCR, which is then cloned into the relevant plasmid in place of the native *L1.LtrB* intron sequence. The primers employed in this amplification introduce the desired intron sequence changes. Software for predicting the sequence of the primers themselves is available online, on a 'pay-per-click' basis at www.sigmaaldrich.co.uk.

2.6.2 Targetron-mediated inactivation of clostridial genes

The first exemplification of Targetron technology within the genus was in *C. perfringens*[82]. As most of the elements of the basic vector available at the time, pACD3, did not function in *Clostridium*, substantive modifications were required. These included provision of a clostridial replicon and selectable marker (from plasmid pJIR750) and replacement of the T7 promoter responsible for transcription of the *Ll.LtrB* group II intron with that of the promoter region of the beta-2 toxin gene (*cpb2*) from a *C. perfringens* type A isolate. The group II intron of the final vector (pJIR750ai) was retargeted to the *plc* gene and introduced into strain 13 by electroporation. Appropriate PCR screening revealed that two of the 38 individual colonies screened contained a mixture of wild-type and mutant (cells in which the intron had inserted into the *plc* gene) using PCR and primers that flanked the predicted insertion site. Plating of the cells from these clones on egg yolk BHI plates revealed that approximately 10% of the colonies formed were alpha-toxin mutants as they were not surrounded by characteristic halos. The presence of the intron in the *plc* gene was confirmed by PCR and shown to be inserted at the predicted site by direct sequencing. Western blotting confirmed that alpha-toxin was not produced, and the mutation was shown to be stably maintained. Since this demonstration, basic Targetrons have been used to generate mutants in two non-pathogenic clostridia, *C. acetobutylicum* and *C. phytofermentans* affecting solvent production and substrate utilisation[83–85].

2.6.3 Positive selection of gene inactivation

Rational alteration of intron sequences is not an exact science, and integration frequencies vary widely between target sites. As a consequence, if insertional inactivation of a target gene does not result in a readily detectable phenotype, substantive screening efforts may need to be expended to isolate the desired mutant. To circumvent this deficiency, Lambowitz and coworkers devised a modified version of the Targetron containing a selectable retrotransposition-activated marker (RAM), the presence of which allows for the positive selection of intron insertion[86,87]. In essence this involved incorporating into the intron an inactive copy of an antibiotic resistance gene which becomes activated during the process of retransposition (insertion of the intron into an alternative target site). The net result is that a fully functional antibiotic resistance gene is co-inserted into the target gene along with the group II intron, an event that can be selected by acquisition of antibiotic resistance[86,87]. The RAM cargo is inserted without affecting intron function in domain IV, a non-structural region formerly occupied by *ltrA*. The antibiotic resistance gene is initially inactive due to the presence of a DNA insertion encoding a group I intron. As self-catalytic splicing of group I introns from RNA transcripts is a strand-specific event, it is orientated in the Targetron plasmid such that splicing does not occur from the mRNA of the transcribed antibiotic resistance gene, but does occur when group II intron RNA is produced. A plasmid incorporating this feature, pACD4K-C, is available from Sigma-Aldrich, where the RAM is based on a modified *kan* gene. Insertion of this modified intron into the target gene is, therefore, selected on the basis of acquisition of resistance to kanamycin.

2.7 The ClosTron: a universal gene knock-out system for clostridia

As the vector used by Chen and coworkers[82] to generate a *plc* mutant of *C. perfringens* lacked a RAM, the detection of the desired insertion was reliant on a combination of PCR screening and a simple phenotypic plate assay. As such convenient assays are available only for a small minority of genes, we elected to construct a more generally applicable system incorporating a clostridial RAM. Such a vector would allow positive selection for integration of group II intron elements into any gene, without recourse to a phenotypic screen, and in clostridial species in which, unlike *C. perfringens*, mutants have seldom or never previously been obtained.

2.7.1 ClosTron development

Previously described RAM elements were based on kanamycin (pACD4K-C: http://www.sigma-genosys.com/targetron/) and trimethroprim[87]. Since neither antibiotic is generally selectable in clostridial species, an alternative clostridial RAM was constructed based on the *ermB* gene from the *Enterococcus faecalis* plasmid pAMβ1. Effective splicing of the group I intron of the *td* gene of phage T4 requires the presence of sequence specific, flanking exon sequences. In the case of the kanamycin and trimethroprim resistance genes, these exon specific sequences could be introduced into the coding sequence without any adverse effect on enzyme function. As such sequences could not be introduced into *ermB* without changing the encoded amino acid sequence, a linker region was inserted between the *ermB* ORF and its ribosome-binding site[88] into which the necessary exon sequences were incorporated. The additional 12 amino acids added to the N-terminus of the ErmB protein were shown to have no discernible effect on function[88]. During the course of the development of the new clostridial RAM, it transpired that transcription from the natural promoter of the *ermB* gene was insufficient to confer erythromycin resistance on the host cells when present as a single copy chromosome insertion[88]. The promoter was, therefore, replaced with the more efficient *thlA* promoter isolated from *C. acetobutylicum* ATCC 824[89]. To control production of group II intron RNA, the final clostridial vector, pMTL007, carried a clostridial promoter (*fac*) from which transcription could be induced by addition of exogenous IPTG in much the same way as the promoter it replaced, T7 (Fig. 2.3).

An important feature of the design was to ensure that the replicon used replicated inefficiently in the clostridial host, and was rapidly lost along with *ltrA*. The continued presence of LtrA in the cell will, in those instances where the insertion of the intron is in the sense strand, bring about splicing of the intron from the mRNA. In this case, insertion of the intron may not be mutagenic. Moreover, LtrA needs to be removed to ensure no additional insertions occur. For this purpose, plasmid pMTL007 was based on the replicon of the plasmid pCB102, as it is lost, albeit with different rapidity, relatively quickly from all of the clostridial species thus far tested. It is most unstable in *C. difficile*[15], where plasmid loss is ensured by simply restreaking the integrant clone. Plasmid segregation is measured by loss of resistance to thiamphenicol.

2.7.2 The prototype ClosTron system

Validation of the constructed ClosTron plasmid pMTL007 was initially undertaken in three different clostridial species, namely *C. difficile, C. acetobutylicum* and *C. sporogenes*[88], targeting two genes with easily-screenable phenotypes, *spo0A* and *pyrF*. The former generated strains defective in sporulation, while the latter resulted in strains with a growth requirement for the supply of exogenous uracil. The changes necessary

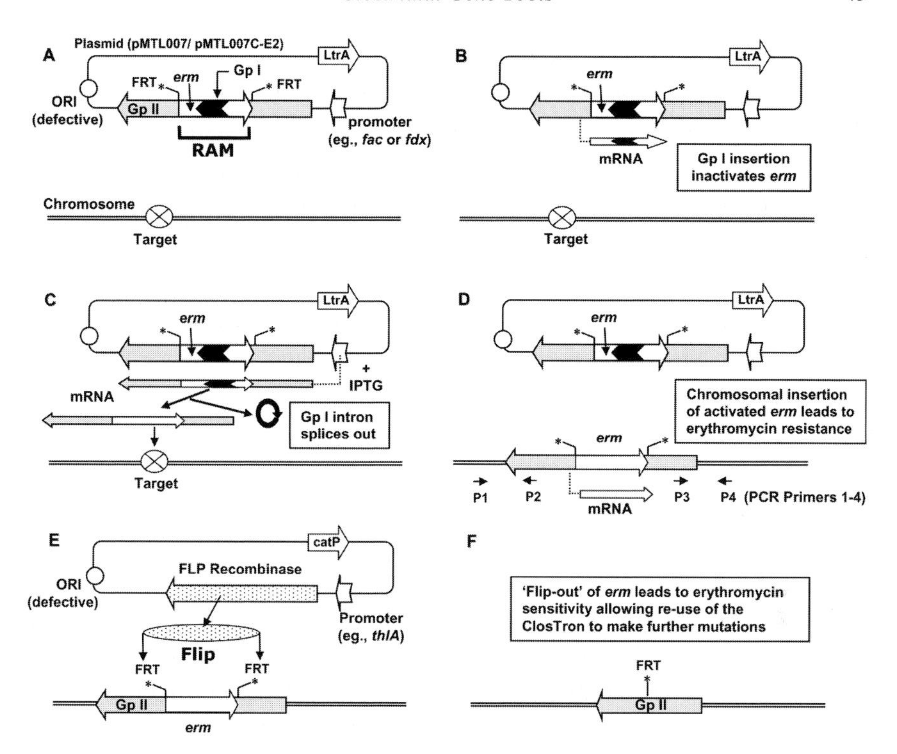

Fig. 2.3: ClosTron mutagenesis. (A) An appropriate ClosTron plasmid is first transferred into the clostridial host. The relative orientations of the group II intron (Gp II) and RAM element (comprising an inactivated *ermB* marker containing group I (Gp I) intron) are crucial. (B) The plasmid-borne *ermB* gene is interrupted by a DNA segment encoding a Gp I intron and, therefore, does not confer erythromycin resistance. The Gp I intron cannot self-splice when *ermB* is expressed, as it is in the reverse orientation. (C) When the Gp II intron RNA is transcribed, the Gp I intron is in the correct orientation, and self-splices. The Gp II intron RNA undergoes retrotransposition into the chromosomal target site. (D) Upon insertion into the chromosome, the *ermB* is now functional, and integrant cells can be selected on the basis of acquisition of erythromycin resistance. Insertion into the intended site can be demonstrated using a combination of primers, e.g. primers P1 + P2, P3 + P4 and P1 + P4. (E) Following loss of pMTL007C-E2, a second plasmid is introduced carrying an *flp* gene under the control of a constitutive promoter, e.g. the *thlA* promoter[89]. (F) Following growth for 10–20 generations in the absence of erythromycin replica plating is used to identify cells that have become sensitive to the antibiotic due to 'flip-out' of *ermB*. Such cells are then cultured in media lacking thiamphenicol to isolate clones cured of the *flp*-bearing plasmid. The ClosTron may now be used again to make a second mutation[90].

to retarget individual plasmids were identified using the web-based computer algorithm at www.sigmaaldrich.com.

Retargeted plasmids were introduced into the appropriate clostridial strain by either electroporation (*C. acetobutylicum*) or conjugation from *E. coli* CA434 (*C. difficile* and *C. sporogenes*) and transfer of the plasmid initially selected on the basis of acquisition of thiamphenicol resistance. Thereafter, cells were restreaked onto media supplemented with erythromycin, but lacking thiamphenicol. To screen for the successful insertion of the intron into the correct target gene, DNA prepared from erythromycin resistant clones was screened by PCR. Screening is best accomplished using three primers (Fig. 2.3). Pair P1 and P2 flank the site of insertion of the group II inton + *ermB* fragment. Such primers generate a relatively small fragment with the wild-type DNA, but result in a fragment 1,800 bp larger in the case of mutants. Primers P1 and P3 (EBS Universal, specific to the element itself, and known to be an effective primer) were used to demonstrate that the element has inserted into the correct gene.

All of the genes targeted were inactivated with high efficiency at the first attempt. Subsequently, several other genes were successfully inactivated, including four in *C. botulinum*[88]. Typically, in each experiment, thousands of erythromycin resistant colonies were obtained, which in every case may be assumed to have been a consequence of insertion of the group II intron and *ermB* into the chromosome. In general, the majority of these integration events took place at the intended target site. Sequencing demonstrated that the insertion point of every mutant was precisely as intended. No evidence was obtained for multiple insertions of the element, and in the case of the three *pyrF* mutants, no evidence of reversion to wild-type uracil prototrophy was apparent. This was tested by plating out undiluted, late exponential cultures of each mutant onto minimal media lacking uracil. As predicted, the mutants are, therefore, extremely stable.

Throughout validation, it only proved necessary to screen four erythromycin colonies for insertion of the element into the correct locus. At least 25% of the clones, if not all, were found to contain the desired insertion. If none of the clones proved positive, an agar plate carrying at least 100 colonies was flooded with media and DNA prepared from the resultant pool of cells. Screening of this pool of DNA with appropriate primers invariably generated a PCR product of the expected size, indicating that the mutant was present at a low frequency. The mutant clone(s) present could then be isolated by more extensive screening.

2.7.3 ClosTron procedure refinements

During the course of validation of the system, a number of refinements that would benefit mutant generation became apparent which have subsequently led to the enhancement of the system[90]. The construction of the new series of plasmids is based on our modular vector format[16].

2.7.3.1 *More accessible retargeted intron design*

Retargeting of the Ll.LtrB intron is accomplished by incorporating complementary sequences to those present in the target gene identified using a computer algorithm[81]. Practically, this entails accessing the Sigma-Aldrich TargeTron Design Site (http://www.sigma-genosys.com/targetron) provided with purchase of the TargeTron Gene Knockout System kit. To maximise the ease and accessibility of ClosTron mutagenesis, we have written an intron design tool based on the data and algorithm described by Perutka and coworkers[81], which is available to the research community free-of-charge via a publicly-accessible website (http://clostron.com).

2.7.3.2 *Conventional intron retargeting*

Conventional retargeting of Targetron/ClosTron plasmids is reliant on substitution of a small 350 bp fragment between the *Hin*dIII and *Bsr*GI restriction sites incorporating the desired changes. In pre-existing plasmids, such as pMTL007, there is no phenotypic screen for detection of this substitution. This has been addressed in our new modular series of plasmids through the incorporation of a small 'stuffer' fragment, including a *lacZα* ORF, which is replaced when the intron is retargeted (Fig. 2.4). Colonies containing retargeted modular ClosTron plasmids are, therefore, no longer blue on plates supplemented with XGal. The presence of an insert of the correct size can also be verified by colony

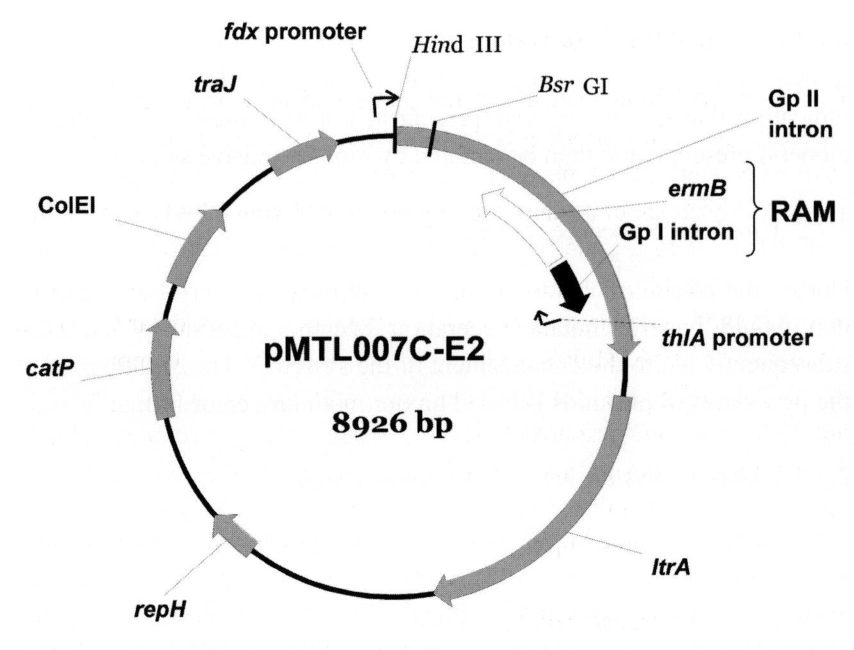

Fig. 2.4: The ClosTron plasmid pMTL007C-E2. The second generation ClosTron plasmid pMTL007C-E2 uses the strong *fdx* promoter from *C. sporogenes* to direct expression of the Gp II intron, and contains FRT sites flanking the RAM to facilitate FLP-mediated marker removal. The plasmid also features a *lacZ'* stuffer sequence that is replaced during intron retargeting, allowing clones containing successfully retargeted plasmids to be identified by blue / white screening, restriction analysis or PCR.

PCR using primers spofdx-seq-F1 and pMTL007-R1, which gives a product of 440 bp from the parental plasmid, or 548 bp after retargeting Finally, restriction analysis may also be employed, as *Sac*I and/or *Bgl*II cut twice in all modular ClosTron plasmids prior to retargeting, and only once after retargeting.

In addition to more effective detection of the successful cloning of the retargeted region, we have also addressed issues related to its initial generation. This is conventionally achieved by SOE PCR using a template supplied in the Sigma-Aldrich TargeTron kit. As this is an exhaustible reagent, we constructed two plasmids (pMTL20IT1 and pMTL20IT2) which can be mixed together and then used as an alternative template for SOE PCR.

2.7.3.3 *Streamlined intron retargeting*

To reduce the labour intensity of the process of plasmid retargeting, and make the procedure more accessible to the expert and novice alike, we have advocated the adoption of a synthetic route to plasmid generation. We take advantage of modern technology by ordering ClosTron plasmids ready-targeted to a gene of choice from an appropriate DNA synthesis company (see http://clostron.com for details). Each intron targeting region is synthesised and cloned by the supplier into a modular ClosTron plasmid such as pMTL007C-E2 (Fig. 2.4). This approach compares very favourably with the conventional approach, in terms of cost (see http://clostron.com for current pricing), labour-intensity (requiring just a few minutes of design and no laboratory work for the researcher) and time from start to finish. Depending on the company utilised, retargeted plasmids are delivered approximately two weeks after ordering.

2.7.3.4 *Plasmid transfer and selection*

The original ClosTron plasmid pMTL007 is based on the *C. butyricum* plasmid pCB102 and utilises the *catP* gene as its selectable marker. Whilst the pCB102 replicon is suitable for transfer to many *Clostridium* spp. and strains, in some instances it is not appropriate, e.g. *Clostridium sordelli*. Moreover, in other cases, resistance to thiamphenicol cannot be used, e.g. *Clostridium beijerinckii*. Accordingly, using modular plasmid components, we have constructed a series of variants incorporating different replicons and markers, with which the range of ClosTron plasmids are extended[90]. In at least one instance, this extension of range was accomplished by using an alternative antibiotic. Thus, whilst hypervirulent ribotype 027 *C. difficile* strains are resistant to erythromycin, by a mechanism other than possession of an *ermB* gene, they retain sensitivity to lincomycin. The applicability of ClosTron mutagenesis to *Clostridium difficile* R20291 (the B1/NAP1/027 strain responsible for the devastating outbreaks between 2003 and 2005 at the UK Stoke Mandeville hospital) was therefore demonstrated through the knock-out of *spo0A* using lincomycin to select the relevant intron insertion[90].

2.7.3.5 *Marker recycling for multiple insertions*

When pMTL007 has been used to generate a mutant, the presence of the *ermB* gene in the target gene means that the ClosTron cannot be used to make a second mutation. To circumvent this deficiency, we constructed an *ermB* RAM flanked by directly repeated FLP Recognition Target (FRT) sites and incorporated it into a new modular ClosTron plasmid, pMTL007C-E2 (Fig. 2.4)[90]. We then generated a retargeted plasmid (using a previously validated retargeting region)[88] to isolate insertional mutants in the *agrA* gene of *Clostridium acetobutylicum* ATCC 824. A plasmid was then constructed (pMTL85151-PPS-flp3) expressing FLP recombinase and introduced into the *agrA* mutant. The resultant transformant was sub-cultured imposing thiamphenicol selection for pMTL85151-PPS-flp3, and colonies screened for loss of erythromycin-resistance due FLP:FRT-mediated excision of *ermB*. Having confirmed the successful excision of *ermB* from a randomly picked erythromycin sensitive clone by PCR and nucleotide sequencing, we were then able to use a retargeted pMTL007C-E2 plasmid to generate a second mutation in the *spo0A* gene, selecting for intron insertion on the basis of acquisition of erythromycin resistance[90].

2.7.3.6 *Delivery of heterologous 'Cargo' sequence*

Previous studies with *C. perfringens* have demonstrated that the Ll.LtrB intron can be used as a vector for inserting transgenes into the genome[91,92]. Such genes may be placed in domain IV of the intron, a region which in the ClosTron is already occupied by the *ermB* RAM. To evaluate the size of DNA that could be delivered in addition to the *ermB* RAM, we undertook a systematic study using fragments of phage lambda DNA (1.0, 2.0 or 2.3 kb) and *Clostridium sporogenes* NCIMB 10696 as the host organism[90]. Each lambda fragment was inserted into the *Sal*I site in domain IV of the intron of pMTL007C-E2. Each intron derivative was retargeted to the previously validated target site in the *pyrF* gene. We did not observe any integrant colonies using the intron derivatives with either of the two larger fragments but were reproducibly able to obtain clones at a frequency of 5×10^{-8} integrants per cell in which the intron contained the 1.0 kb fragment. Whilst the frequency of isolation was 104-fold lower than the intron containing only the *ermB* RAM, it nonetheless demonstrated the feasibility of the selectable delivery of small transgenes into the genome[90].

2.7.3.7 *Constitutive expression of group II RNA*

During the course of validation of the pMTL007 system[88], it became apparent that the regulated expression of group II intron RNA was not necessary. Thus, constitutive expression from *fac*, achieved by the introduction of a frame-shift into the *lacI* gene on pMTL007, did not reduce the number of mutants obtained in any of the clostridial strains tested. Indeed, in the organism in which *fac* was under the tightest regulatory control, *C. sporogenes*, the total number of mutants generated was higher, by an order of magnitude, following the introduction of a frame-shift into the *lacI* gene. As a consequence, our new modular ClosTron plasmids[90] no longer incorporate the *lacI* gene, relying on constitutive expression of the group II intron RNA.

2.8 Conclusions

In recent years the development of gene tools for the genus *Clostridium* has intensified, particularly within the arena of genomic manipulation. Formerly, appreciable mutant generation was confined to *C. perfringens* strain 13, in part a reflection of its intrinsic high frequency of transformation. Thus, in *C. perfringens* mutant generation was on a scale unmatched by any other clostridial species. The development of the ClosTron and related TargeTron methodologies has gone some way to redressing the balance, allowing the rapid creation of stable mutations[82–84–85,93–98]. Moreover, with the recent adaptations made to the system (http://www.clostron.com)[90], in terms of minimising the labour-intensity and maximising the accessibility of the mutagenesis method, the addition of the facility to make multiple mutations and the creation and distribution of a modular vector system[16], the pace of developments is likely to intensify. In parallel to the developments made in the ability to make directed mutations for reverse genetic studies, there is now light at the end of the tunnel in terms of tools for random mutant generation that may be applied in forward genetic approaches. Thus, the demonstration of the applicability of both transposome technology and a *mariner*-based transposon in *C. perfringens* and *C. difficile,* respectively[48–50], represent a considerable step forward in the armoury available to the clostridial molecular biologist.

Despite the progress made, there is still some way to go before clostridial researchers can boast the degree of sophistication available to those researchers studying *Bacillus*. Single cross-over mutants, ClosTron/TargeTron mutants and transposon generated mutants are insertional mutants and, therefore, suffer from the spectre of polar effects. The development of rapid and reliable methods for creating in-frame deletions will be pivotal. Similarly, although ClosTron and TargeTron elements may be used to deliver *trans* genes to the chromosome, the size of DNA that can be inserted is limited. More effective methods which can deliver large segments of DNA are clearly required, particularly for the successful application of synthetic biology approaches to enhancing the properties of commercially useful strains. We would envisage that the elements such as the ClosTron will always have a role to play when the rapid creation of an insertional mutant is required, or when small *trans* genes need to be delivered to the chromosome[90], but that for more precise or more ambitious modifications, alternative methods will be brought to bear. Such methods have already reached an advanced stage of development in our laboratory.

Acknowledgement

This work was supported by the BBSRC (grants BB/E021271/1, BD/F003390/1 and BB/G016224/1), the MRC (G0601176) and Morvus Technology Ltd.

References

1. Young M. *et al.*, *FEMS Microbiol. Rev.* 63 (1989).
2. Rood J. I. and Cole S. T., *Microbiol. Rev.* 55 (1991).

3. Minton N. P. *et al.*, in *The Clostridia and Biotechnology*, Woods D. R., Ed. (Butterworth-Heinemann, Stoneham, 1993), pp. 119–150.
4. Mauchline M. L. *et al.*, in *Manual of Industrial Microbiology and Biotechnology,* Demain A. L. and Davies J. E., Eds (ASM Press, Washington DC, 1999), pp. 475–492.
5. Davis I. *et al.*, in *Handbook on Clostridia,* Duerre P., Ed. (CRC Press, Boca Raton, 2005), pp. 37–52.
6. Young D. I. *et al.*, *Methods Microbiol.* 29 (1999).
7. Tyurin M.V. *et al., Appl. Environ. Microbiol.* 70 (2004).
8. Trieu-Cuot P. *et al.*, *FEMS Microbiol. Lett.* 48 (1987).
9. Bertram J. and Dürre P., *Arch. Microbiol.* 151 (1989).
10. Woolley R. C. *et al.*, *Plasmid.* 22 (1989).
11. Williams, D. R. *et al.*, *J. Gen. Microbiol.* 136 (1990).
12. Mermelstein L. D. and Papoutsakis E. T., *Appl. Environ. Microbiol.* 59 (1993).
13. Jennert K. C. *et al.*, *Microbiol.* 146 (2000).
14. Davis T. O. *et al., J. Mol. Microbiol. Biotechnol.* 2 (2000)
15. Purdy D. *et al.*, *Mol. Microbiol.* 46 (2002).
16. Heap T. J. *et al., J. Microbiol. Methods* 78 (2009).
17. Clewell D. B., in *Bacterial Genomes: Physical Structure and Analysis*, deBruijn F. J. *et al.*, Eds (Chapman and Hall, International Thomson Publishing, Thomson Science, New York, 1998), pp. 130–139.
18. Ionesco H., *Ann. Microbiol. (Paris)* 131A(2) (1980).
19. Smith C. J. *et al., Antimicrob. Agents Chemother.* 19 (1981).
20. Wüst J. and Hardegger U., *Antimicrob. Agents Chemother.* 23 (1983).
21. Mullany P. *et al.*, *J. Gen. Microbiol.* 136 (1990).
22. Roberts A. P. *et al.*, *Microbiol.* 147 (2001).
23. Sebaihia M. *et al.*, *Nat. Genet.* 38 (2006).
24. Lin W. J. and. Johnson E. A., *Appl. Env. Microbiol.* 57 (1991).
25. Kaufmann P. et al., Syst. Appl. Microbiol. 19 (1996).
26. Lyristis M. *et al.*, *Mol. Microbiol.* 12 (1994).
27. Banu S. *et al.*, *Mol. Microbiol.* 35 (2000).
28. Lin W. J. and Johnson E. A., *Appl. Environ. Microbiol.* 61 (1995).
29. Mullany P. *et al., FEMS Microbiol. Lett.* 79 (1991).
30. Wang H. *et al., FEMS Microbiol. Lett.* 192 (2000).
31. Roberts A. P. *et al.*, *J. Microbiol. Methods* 55 (2003).
32. Hussain H. A., *J. Med. Microbiol.* 54 (2005).
33. Bertram J. *et al.*, *Arch. Microbiol.* 153 (1990).
34. Mattsson D. M. and Rogers P., *J. Ind. Microbiol.* 13 (1994).
35. Babb B. L. *et al.*, *FEMS Microbiol. Lett.* 14 (1993).
36. Sauer U. and Dürre P., *FEMS Microbiol. Lett.* 114 (1993).
37. Collett H.J., *et al.*, *Anaerob.* 3 (1997).
38. Kashket E.R. and Cao Z.Y., *Appl. Environ. Microbiol.* 59 (1993).

39. Kashket E.R. and Cao Z.Y., *FEMS Microbiol. Rev.* 17 (1995).
40. Liyanage H. *et al., J. Mol. Microbiol. Biotechnol.* 2 (2000).
41. Evans V. J. *et al., Appl. Environ. Microbiol.* 64 (1998).
42. Awad M. M. and Rood J. I., *Microb. Pathog.* 22 (1997).
43. Briolat V. and Reysset G., *J. Bacteriol.* 184 (2002).
44. Bae T. *et al., P. Natl. Acad. Sci. USA* 101 (2004).
45. Cao M. *et al., Appl. Environ. Microbiol.* 73 (2007).
46. Le Breton Y. *et al., Appl. Environ. Microbiol.* 72 (2006).
47. Wilson A. *et al., J. Microbiol. Methods* 71 (2007).
48. Lanckriet A. *et al., Appl. Environ. Microbiol.* 75 (2009).
49. Vidal J. E. *et al., PLoS One.* 14;4(7):e6232 (2009).
50. Cartman S. T. and Minton N. P., *Appl. Environ. Microbiol.* 76 (2010).
51. Gao L.Y. *et al., Infect. Immun.* 71, 922 (2003).
52. Maier T. M. *et al., Appl. Environ. Microbiol.* 72 (2006).
53. Lampe D. J. *et al., Genetics* 149 (1998).
54. Dupuy B. and Matamouros S., *Res. Microbiol.* 157 (2006).
55. Dupuy B. and Sonenshein A. L., *Mol Microbiol.* 27 (1998).
56. Kaniga K. *et al., Gene.* 109 (1991).
57. Shimizu T. *et al., J. Bacteriol.* 176 (1994).
58. Awad M. M *et al., Mol. Microbiol.*, 15 (1995).
59. Awad M. M. *et al., Microb. Pathog.* 28 (2000).
60. Sarker M. R. *et al., Mol. Microbiol.* 33 (1999). Erratum in: *Mol. Microbiol.* 35, 249 (2000).
61. Keyburn A. L. *et al., Infect. Immun.* 74 (2006).
62. Kennedy C. L. *et al., Mol. Microbiol.* 57 (2005).
63. Awad M. M. *et al., Infect. Immun.* 69 (2001).
64. Awad M. M. and Rood J I., *J. Bacteriol.* 184 (2002).
65. Ohtani K. *et al., FEMS Microbiol. Lett.* 209 (2002).
66. O'Brien D. K and Melville S. B., *Infect. Immun.* 72 (2004).
67. Varga J. J. *et al., Mol. Microbiol.* 62 (2006).
68. Liyanage H. *et al., Appl. Environ. Microbiol.* 67 (2001).
69. Wilkinson S. R. and Young M., *Microbiol.* 140 (2001).
70. O'Connor J. R. *et al., Mol. Microbiol.* 61 (2006).
71. Dineen S. S. *et al., Mol. Microbiol.* 66 (2007).
72. Heap J. T. *et al.*, in *Clostridia: Molecular biology in the post-genomic era*, Brüggemann H. and Gottschalk G., Eds (Caister Academic Press, Norfolk, UK, 2008), pp. 179–198.
73. Reid S. J. *et al., Microbiol.* 145 (1999).
74. Green E. M. and Bennett G. N, *Appl. Biochem. Biotechnol.* 57/58 (1996).
75. Green E. M. *et al., Microbiol.* 142 (1996).
76. Nair R. V. *et al., J. Bacteriol.* 181 (1999).
77. Harris L. M. *et al., J. Bacteriol.* 184 (2002).

78. Fabret C. *et al.*, *Mol. Microbiol.* 46 (2002).
79. Lambowitz A. M. and Zimmerly S., *Annu Rev Genetics* 38 (2004).
80. Mohr G. *et al.*, *Genes Dev.* 14 (2000).
81. Perutka J. *et al.*, *J. Mol. Biol.* 336 (2004).
82. Chen Y. *et al.*, *Appl. Environ. Microbiol.* 71 (2005).
83. Shao L. *et al.*, *Cell Res.* 17 (2007).
84. Jiang Y. *et al.*, *Metab Eng.* 11 (2009).
85. Wilson D. B., *Mol. Microbiol.* 74 (2009).
86. Karberg M. *et al.*, *Nat. Biotechnol.* 19 (2001).
87. Zhong J. *et al.*, *Nucleic Acids Res.* 31 (2003).
88. Heap J. T., *Microbiol. Methods* 70 (2007).
89. Girbal L. *et al.*, *Appl. Environ. Microbiol.* 69 (2003).
90. Heap J. T. *et al.*, *J. Microbiol. Methods.* 80 (2010).
91. Plante I. and. Cousineau B, *RNA* 12 (2006).
92. Chen Y. *et al.*, *Plasmid.* 58 (2007).
93. Bradshaw M. *et al.*, *Appl. Environ. Microbiol.* 76 (2010).
94. Burns D. A. *et al.*, *J Bacteriol.* 192 (2010).
95. Emerson J. E. *et al.*, *Mol. Microbiol.* 74 (2009).
96. Kirby J. M. *et al.*, *J. Biol. Chem.* 284 (2009).
97. Twine S. M. *et al.*, *J. Bacteriol.* 191, 7050 (2009).
98. Underwood S. *et al.*, *J. Bacteriol.* 191, 7296 (2009).

CHAPTER 3

SUPPORTING SYSTEMS BIOLOGY OF *CLOSTRIDIUM ACETOBUTYLICUM* BY PROTEOME ANALYSIS

H. JANSSEN

University of Illinois at Urbana-Champaign
Department of Food Science and Human Nutrition
1207 West Gregory Drive
Urbana, Illinois 61801, USA

T. FIEDLER

Institute of Medical Microbiology, Virology, and Hygiene
Rostock University Hospital
Schillingallee 70, 18057 Rostock, Germany

K. SCHWARZ

Clostridia Research Group, Centre for Biomolecular Sciences
University Park, University of Nottingham
Nottingham NG7 2RD, UK

H. BAHL, R.-J. FISCHER

Institute of Biological Sciences
University of Rostock
Albert-Einstein-Strasse 3, 18059 Rostock, Germany

Clostridium acetobutylicum is a strictly anaerobic, Gram-positive bacterium and is characterized by its fermentation products acetone and butanol. Its potential for the biotechnological production of these bulk chemicals initiated intensive research activities and *C. acetobutylicum* became a model organism for apathogenic clostridia. Thus, it is not surprising that this organism is also an attractive candidate for systems biology studies. Several of the necessary "-omic" technologies, including proteomics, have been applied to *C. acetobutylicum*. However, in relation to other microorganisms, the knowledge of the proteome of *C. acetobutylicum* is limited.

In this chapter the development and the state of the art of proteome analysis of *C. acetobutylicum* are reflected upon, including stress proteomes and first results of the secretome. Special emphasis is placed on the changes of the proteome related to the metabolic shift of *C. acetobutylicum* from acid to solvent production. A main objective of this chapter is a detailed and integrated comparison of existing proteome and transcriptome data, and the limitations of both technologies are discussed. Finally, future perspectives of the proteomics in systems biology are reflected upon.

3.1 Introduction

The beginning of the era of "systems biology" is based on an increasing number of "-omic" technologies such as genomics, transcriptomics, proteomics, metabolomics, interactomics, complexomics, localomics, glycomics, or lipidomics[1,2]. They all have in common the generation of large data sets, representing snapshots of different aspects of a living organism. Each technology delivers just one piece of the puzzle of a comprehensive description of a biological system. Usually, data are mathematically integrated by specialists in bioinformatics groups to develop models of the whole organism, or at least of a part of its metabolism[3]. Progress in this field regarding *C. acetobutylicum* is reported elsewhere in this book.

The interest in *C. acetobutylicum* has historically arisen due to its ability to produce the solvents butanol and acetone that are required as industrial bulk chemicals. This property led to extensive biotechnological exploitation of this obligate anaerobic bacterium and triggered a variety of research projects. Currently, a field enjoying renewed attention is the use of *C. acetobutylicum* for the biotechnological production of butanol as a biofuel. This potential makes the organism and its metabolism valuable from both an ecological and economical point of view[4,5,6]. Unquestionably, today *C. acetobutylicum* is one of the best investigated apathogenic clostridial species and is considered a model system for this group of bacteria. This was substantially supported by recent breakthroughs in the development of tools enabling its genetic manipulation[7,8]. Consequently, this bacterium was an obvious choice for systems biology projects like SysMO-COSMIC1 (http://www.sysmo.net).

The relative importance of different "-omic" techniques used in systems biology approaches varies. This is not surprising because individual weightings and perceptions are changing due to methodological improvements and technical developments. Additionally, it cannot be denied that the expertise and the technological equipment of individual laboratories play a prominent role.

Proteome analysis is one of the oldest "-omic" techniques, relying on the pioneering development of the two-dimensional electrophoresis (2-DE) by O'Farrel[9]. However, until today only limited proteome data of *C. acetobutylicum* were available in comparison to other microorganisms and all peer-reviewed publications dealt with the soluble protein fraction of the cytosol.

2-DE studies of *C. acetobutylicum* started in the 1980s, long before the genome sequence was annotated. Figure 3.1 shows 2-DE gels documented by Bahl (1983) which likely represent the first proteome study of *C. acetobutylicum* cells[10].

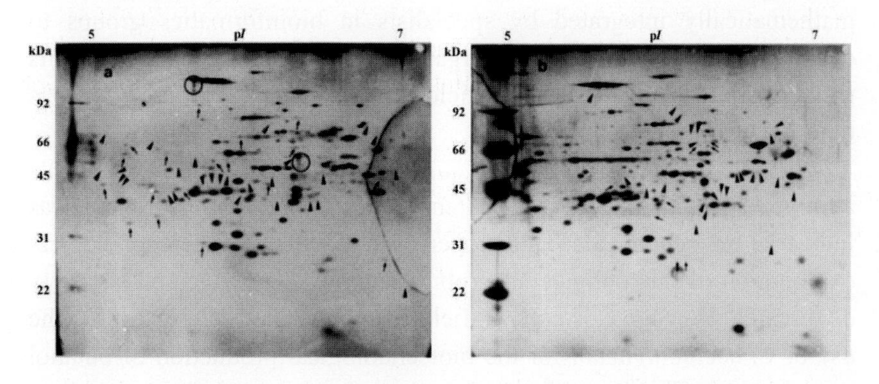

Fig. 3.1: First 2-DE gels of cytosolic proteins of steady-state chemostat cells of *C. acetobutylicum*[10]. The left gel (a) shows silver stained proteins (40 µg) of acidogenic cells and the right gel (b) proteins of solventogenic cells, respectively. Filled triangles mark absent protein spots and black arrows highlight lower abundant protein spots in comparison to the other steady-state conditions.

Other studies focused on responses to stresses like heat, butanol, or oxygen[11–14] and the onset of the solvent production or the comparison of acidogenic and solventogenic cells[15–17]. Starting with the report of Sullivan and Bennett (2006), investigations of mutant strains gained in importance[17–19].

In the recent past transcriptome data of *C. acetobutylicum* cells have gradually been published[20–27]. Although these data contribute considerably to the understanding of different cellular processes, measurement of mRNA amounts alone is insufficient for comprehensive imaging of a bacterium as a system, because mRNA quantities do not necessarily reflect abundances or even biological activities of the corresponding proteins. For this reason proteome analysis is a good combinatorial tool with which to visualize real protein amounts and furthermore can be useful to indicate posttranscriptional regulation processes and posttranslational modifications[28,29].

This chapter reflects the development of proteome analyses of *C. acetobutylicum*. Firstly, outcomes of so-called stress proteomes are summarized, followed by a paragraph dealing with the knowledge concerning changes of protein patterns accompanying the metabolic switch from acidogenesis to solventogenesis. These investigations are mostly based on batch cultures. This type of cultivation is characterized by permanently changing extracellular and intracellular conditions which are discussed to be less than perfect for the requirements of modellers[16,30].

A kind of breakthrough might be recent studies analyzing cells growing continuously in chemostat cultures. Here, the change of a single parameter in the form of the external pH triggers *C. acetobutylicum* to switch its metabolism from acidogenesis to solventogenesis or *vice versa*. All other parameters were kept constant and cells grow exponentially representing the acidogenic or the solventogenic growth phase. Data of such chemostat cultures of *C. acetobutylicum* have been used to create a computational model of the pH-dependent metabolic shift in *C. acetobutylicum*[30].

A separate paragraph discusses the changes of the proteome of acidogenic cells in comparison to solventogenic cells of batch and of continuous cultures. Beyond that, secretome results are represented giving first insights into the extracellular protein fraction of *C. acetobutylicum*. Finally, exemplary proteome data are compared and correlated with microarray data and limitations of the methods are discussed. The chapter closes with an outlook describing perspectives of proteome data in systems biology projects.

3.2 Stress proteomes of *C. acetobutylicum*

In 1988, Terracciano *et al.* published first images resulting from investigations of stress proteins of *C. acetobutylicum*[11]. Synthesis rates of cytosolic proteins were documented using pulse-labeling with [^{3}H]leucin to unravel the effects of heat shock, solvent (butanol), and oxygen stress. In heat shocked cells an upregulation of eight proteins and a downregulation of two proteins could be demonstrated. Overlapping, seven proteins showed a similar induction in reaction to heat and butanol treatment, and two of them seemed also be induced after the exposure of the cells to air. This finding indicated the existence of general stress proteins in *C. acetobutylicum*.

The heat shock response data were confirmed in a refined experimental set-up by Pich *et al.* (1990)[12]. Since the genome sequence of *C. acetobutylicum* was not known, immune-precipitation experiments have been applied to show similarities to the known stress proteins DnaK[11,12] and Hsp67[12] of *E. coli*.

Annotation of the genome sequence of *C. acetobutylicum*[31] largely facilitated the identification of other protein spots with significantly enhanced abundances in response to heat shock, such as DnaK, GroEL, GroES, Hsp18, Hsp21L, and Hsp21R[13,14].

Interestingly, Hsp21 appeared in 2-DE gels in two spots named Hsp21L (A) and Hsp21R (B) (Fig. 3.2).

It transpired that the Hsp21 proteins are encoded by the two open reading frames *ca_c3598* and *ca_c3597*, the gene products of which revealed striking similarities to rubrerythrins[31]. In comparison to "common" rubrerythrin proteins (Rbr) of other bacteria, the Hsp21/Rbr proteins of *C. acetobutylicum* revealed a "reverse" composition of their two functional domains[32]. These rubrerythrin-like proteins consist of a rubredoxin-like FeS_4 domain at the N-terminus and the ferritin-like

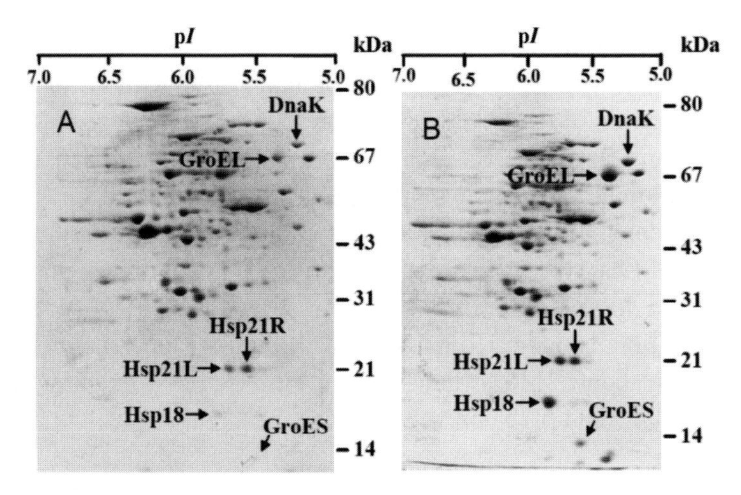

Fig. 3.2: Cytosolic proteins of *C. acetobutylicum* before (A) and after (B) a heat shock[14]. For a heat shock the temperature was shifted for 30 minutes from 37°C to 42°C. Protein extracts were separated by 2-DE and visualized by colloidal coomassie staining. Selected heat shock proteins are highlighted.

di-iron domain (rubrerythrin domain) at the C-terminus. They were denoted as revRbr's and named Rbr3A and Rbr3B because two "normal" Rbr proteins had already been annotated in the genome[14].

The genes *rbr3A* and *rbr3B* are organized in a tandem genetic architecture forming a bicistronic operon. Noticeably, the resulting proteins differ in only one amino acid residue (99% identical amino acid residues). Because of the same theoretical isoelectric points (p*I* value of pH 5.5) and nearly identical calculated molecular weights, they should not be distinguishable on 2-DE gels.

Nevertheless, the proteins Hsp21 or Rbr3A/3B appeared in at least two spots and could not be assigned to Rbr3A or Rbr3B by means of MALDI-TOF[12–14]. However, the appearance of two spots might indicate some unknown posttranslational protein modification.

Results on the protein level underscored a function of Hsp21 or Rbr3A/3B as a general stress protein in *C. acetobutylicum* playing a role in the responses to a heat shock, after oxidative stress and in cells growing under microoxic conditions. In summary, proteome analysis was a valuable tool with which to demonstrate that Hsp21 (Rbr3A/3B) is not only a heat shock protein but also a twin pair of general stress proteins.

Consecutive studies in this field included DNA-microarray data and led to the identification of a regulatory protein (PerR) with major impact on the regulation of the oxidative stress response in *C. acetobutylicum*[33]. A comparative proteome analysis of a *perR* deletion strain and a wild-type strain of *C. acetobutylicum* underlined the regulatory function of PerR as a repressor of the reverse rubrerythrins Rbr3A/3B. Furthermore, under the chosen experimental set-up, Rbr3A/3B proteins were present in no less than four spots with horizontal distribution in 2-DE gels (Fig. 3.3). This finding indicated proteins with similar apparent molecular weights of about 21 kDa but different isoelectric points (p*I* values between pH 4.9 and pH 5.3).

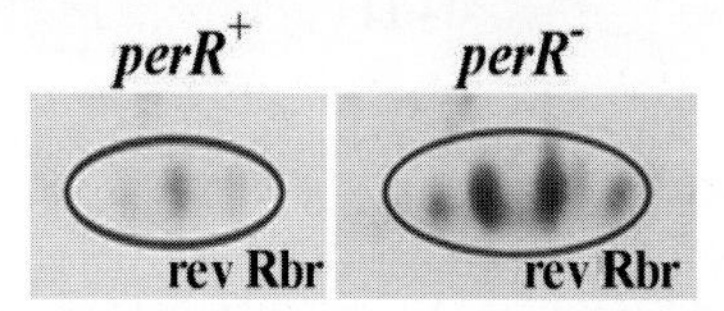

Fig. 3.3: Reverse rubrerythrin proteins (rev Rbr3A/3B, synonymous Hsp21) in a wild-type (*perR*+) and in a PerR deletion (*per*−) strain *of C. acetobutylicum*[33]. Shown are selected spots of 2-DE with protein extracts representing “anaerobiosis”.

Interestingly, a fifth spot of Rbr3A/3B appeared displaying a more than doubled apparent molecular weight of about 55 kDa. It should be mentioned that four other proteins with higher abundances in the *perR* mutant were identified, an A-type flavoprotein (CA_C1027), a flavodoxin encoded by *ca_c2542*, and two unidentified proteins. These findings were directly involved in innovative studies demonstrating that the obligate anaerobic bacterium *C. acetobutylicum* employs an alternative, reductive defence system to relieve oxidative stress, which may well necessitate textbooks being rewritten[33–34].

Another kind of stress investigated with respect to *C. acetobutylicum* is phosphate (P_i) limitation. For this purpose an opposite strategy to the above-mentioned stress experiments was chosen. Cells were grown continuously in a chemostat culture in a synthetic medium under phosphate limiting conditions and, after reaching steady-state conditions, a P_i pulse was applied[35]. 2-DE analysis confirmed increased appearance of the two-component system PhoP/R under P_i limitation indicating its involvement in phosphate-dependent gene regulation. Further investigations of the effects of P_i limitation on the protein composition led to the first standard operating procedure (SOP) for the preparation of intracellular and extracellular proteins and the corresponding protein reference maps mentioned in the following paragraph.

3.3 Sample preparation and proteome reference maps of *C. acetobutylicum*

The studies concerning effects of P_i limitation revealed that for the reliability and reproducibility of 2-DE experiments the reproducible preparation of protein extracts is a crucial prerequisite. Thus, Schwarz *et al.* (2007) developed an SOP for the preparation of soluble cytosolic and soluble extracellular protein samples of *C. acetobutylicum*[36]. Different cell disruption techniques were tested and compared for their applicability in proteome analysis. In brief, a maximum of intact soluble cytosolic proteins could be obtained by sonication of the cells in the presence of the chelator EDTA and PMSF as protease inhibitor, whereas the highest amounts of secreted proteins resulted from a combined ultrafiltration-dialysis procedure (Fig. 3.4).

The first secretome analysis of *C. acetobutylicum* ATCC 824 cells grown in chemostat cultures at pH 5.3 (acidogenesis) and pH 4.5 (solventogenesis) was described by Schwarz (2007)[37]. Figure 3.5 exemplarily illustrates cell-free culture supernatants of cells growing under high phosphate conditions (P_i> 1 mM). In the secretome of solvent-producing cells, 152 spots could be detected. In 140 spots, 51 different proteins were identified by peptide mass fingerprinting (PMF). Six of them did not comprise a signal sequence for transportation over the cytoplasm membrane[38]. Thus, they seemed to belong to typical cytosolic proteins.

Almost half of the identified secreted proteins are not functionally characterized. The majority of the proteins are involved in carbohydrate transport and carbohydrate metabolism or in cell envelope biogenesis.

In comparison, in the secretome of cells producing acids at pH 5.5 more than twice as many spots (343) were detected and the proteins of 282 spots were identified. They represented 116 different proteins.

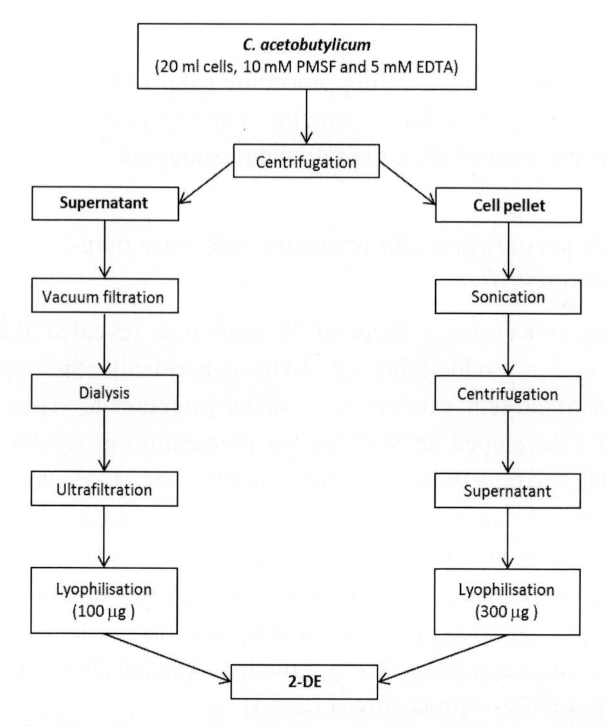

Fig. 3.4: Standard operating procedure for the preparation of samples of extracellular or intracellular proteins of *C. acetobutylicum*[36]. Originally, samples of chemostat cultures were processed and all steps were preferably performed at 4°C. If necessary, protein samples were stored overnight at –20°C.

Strikingly, only 41 (35%) of them could be assigned to the secretome based on the criteria of Desvaux *et al.* (2005)[38].

The reason why the majority of extracellular proteins of acidogenic cells obviously seemed not to be actively secreted remains questionable so far. As the preparation of the supernatants of both cell types exactly followed the SOP it could be speculated that more cells are disrupted during the preparation of the cell-free supernatant at pH 5.5. An explanation might be a specific cell wall or cell membrane component composition of acidogenic cells[37].

In parallel, using the same experimental set-up, the cytosolic phosphate proteome was investigated employing 2-DE and MALDI-TOF

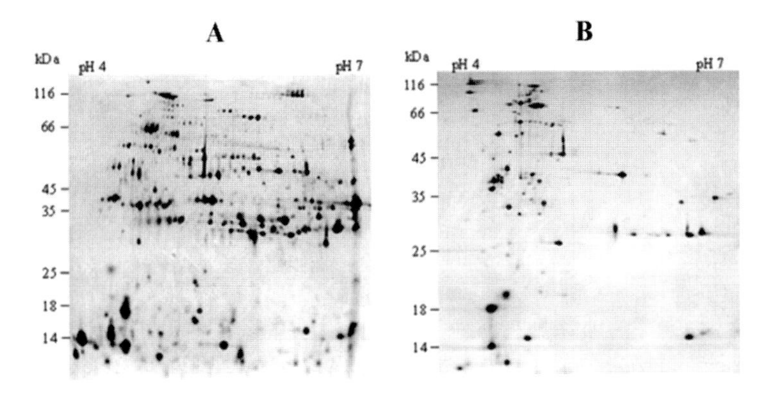

Fig. 3.5: Secretomes of acidogenic (A) and solventogenic (B) cells of *C. acetobutylicum*[37]. Samples were taken out of a phosphate limited chemostat culture at pH 5.5 (A) or at pH 4.5 (B) after a P_i pulse (>1 mM P_i). 2-DE gels contained 100 μg of extracellular protein and were stained by colloidal coomassie.

analyses. Differences in response to P_i pulses were monitored[39] and a phosphate proteome reference map was generated (data not shown). Altogether, 36 proteins showed significantly induced abundances (ratio $\geq$ 2) under phosphate limited growth (P_i< 1 mM, "low P_i") in comparison to growth under conditions of P_i saturation (P_i>1 mM, "high P_i").

Results are summarized in Tables 3.1 and 3.2. Only three proteins were found to be significantly (ratio $\geq$2) upregulated in reaction to P_i limitation at pH 5.5 as well as at pH 4.5. These are the phosphate-dependent response regulator PhoP[35], the stationary phase regulator

Table 3.1: Proteins with increased abundances (ratio "low P_i"/"high P_i" $\geq$2) in a chemostat culture of *C. acetobutylicum* under conditions of phosphate limitation and acidogenic growth at pH 5.5.

ORF	Protein	Ratio	Fraction
cap0129	glycogen binding S/T protein phosphatase I subunit	4.6	extracellular
cac0430	glycerophosphorylphosphodiesteraseGlpQ	2.8	cytosolic
cac0310	regulator AbrB	2.7	cytosolic
cac1700	response regulator PhoP	2.3	cytosolic
cap0168	alpha-amylase	2.1	extracellular
cac1080	uncharacterized surface-associated protein	2.0	extracellular

Table 3.2: Proteins with increased abundances (ratio "low P_i"/"high P_i" ≥2) in a chemostat culture of *C. acetobutylicum* under conditions of phosphate limitation and solventogenic growth at pH 4.5.

ORF	Protein	Ratio[a]	Fraction
cap0058	lipoprotein A (RLPA) like protein	∞	extracellular
cac1700	response regulator PhoP	7.5	cytosolic
cac0816	lipase esterase like protein	6.8	cytosolic
cac0430	glycerphophorylphosphodiesteraseGlpQ	4.4	cytosolic
cac0310	regulator AbrB	4.3	cytosolic
cac1709	phosphate uptake regulator PhoU	3.1	cytosolic

[a]: The sign for non-terminating (∞) indicates that a protein was not detectable at high P_i.

AbrB, and the glycerophosphoryl phosphodiesterase GlpQ. PhoU, a putative phosphate uptake modulator, was found to be significantly induced in solventogenic cells. Most of the other identified proteins showing increased abundances under phosphate limitation are obviously not associated with phosphate metabolism[39].

Recently, two further, more comprehensive, cytoplasmic proteome reference maps of *C. acetobutylicum* were published by Mao *et al.* (2010)[17] and by Janssen *et al.* (2010)[16].

In the first study, acidogenic and solventogenic cells of a batch culture of *C. acetobutylicum* DSM 1731 were analyzed[17]. The resulting cytoplasmic proteome reference map contained 1206 protein spots (p*I* range between pH 4 and pH 11), caused by 564 different identified proteins. This number covers 14.7% of the 3847 of the deduced open reading frames (3671 encoded by the chromosome and 176 residing on the megaplasmid pSOL1) of this organism[31]. For classification purposes, proteins were categorized into 20 functional orthologous groups (COGs)[40]. The largest group (COG class J, 13.1%) consists of proteins involved in biogenesis, whereas the second largest group (12.0%) contained proteins of general function prediction only (COG class R). The third (10.3%) and fourth (6.2%) largest groups include proteins involved in amino acid transport and metabolism (COG class E) and sugar metabolism-related proteins, especially those proteins involved in energy metabolism (COG class C).

The second comprehensive proteome reference map of *C. acetobutylicum* ATCC 824 was published by Janssen *et al.* (2010)[17]. In this work, exponentially growing steady-state cells of a phosphate-limited chemostat culture representing solventogenic metabolism at pH 4.5 and acidogenic metabolism at pH 5.7 were comparatively analyzed. In this case more than 21% of the cytoplasmic proteins were found. However, calculation of the coverage was restricted to proteins with theoretical isoelectric points between pH 4 and pH 7 and molecular weights between 10 and 200 kDa[41]. Protein extracts of acidogenic cells yielded 357 spots with 178 identified proteins, and in solventogenic cells 205 different proteins were represented in 415 spots. Altogether, a considerably smaller fraction of 251 different proteins (6.5%) was identified in comparison to Mao *et al.* (2010) (564 different proteins, respectively 14.7%)[16].

A possible explanation could be that Mao *et al.* (2010)[16] additionally captured the majority of the alkaline proteins because an extended pH range (p*I* between pH 6 and pH 11) was used in the isoelectric focusing (IEF) step (first dimension of 2-DE). Janssen *et al.* (2010)[17] used a narrower pH range of p*I* values between pH 4 and pH 7. In this case the 2-DE gels clearly revealed a "smear" of condensed proteins at the right edge of the gels indicating alkaline proteins which could not be separated during IEF (indicated in Fig. 3.5, the "dark" lane comprising e.g. AdhE1, CtfA, CtfB, and Adc).

Moreover, the protein content of the analyzed samples differed significantly. Janssen *et al.* (2010)[17] used 300 μg lyophilized proteins whereas Mao *et al.* (2010)[16] applied more than three times as much protein (1 mg). Certainly, a higher protein load led to a higher number of detectable protein spots in 2-DE gels. On the other hand, as a drawback it has to be tolerated that protein spots may be "overloaded", implicating imprecise detection of protein abundances. Basically, although it is questionable to which extent the different culture conditions contributed to different numbers of expressed proteins, in both studies same metabolic phases were analyzed and acidogenic or solventogenic cells were investigated. However, in contrast to the growth under steady-state conditions in a chemostat culture in batch cultures environmental and cellular conditions continuously are changing. It can be speculated that for this reason also, a higher number of proteins might be necessarily expressed. In this context it must be noted, that in P_i limited chemostat culture under conditions of steady-state growth the cells do not underlie the natural cell cycle and events like synthesis of granulose or sporulation generally never occur.

Nevertheless, both reference maps enabled new detailed insights into the protein pattern of acidogenic and solventogenic cells of *C. acetobutylicum* under different cultivation conditions and on different media. Documented differences with respect to batch or continuous chemostat cultivation will be ruled out in the following section.

3.4 Effects of the metabolic shift on the proteome of *C. acetobutylicum*

Research interests have focused on the understanding of the regulation of the "metabolic shift" of *C. acetobutylicum*, that is the change of its metabolism from a predominant production of acids (acetate and butyrate) towards the synthesis of the solvents acetone, butanol, and ethanol. Consequently, proteomes of acidogenic and solventogenic cells have been investigated. Terracciano *et al.* (1988)[11] and Pich *et al.* (1990)[12] presented the first 2-DE based proteomic studies comparing cells of batch and of continuous cultures, respectively. Sullivan and Bennett (2006)[18] and Mao *et al.* (2010)[17] published detailed analyses of growth phase-dependent changes in the protein patterns in batch cultures of *C. acetobutylicum* in a complex medium.

Cells growing continuously in a chemostat culture were investigated by Schaffer *et al.* (2002)[15] and Janssen *et al.* (2010)[16]. Continuous cultivation in principle followed the pioneering studies of Bahl *et al.* (1982)[42]. Always a synthetic medium combining a surplus of glucose with growth limiting concentrations of phosphate (<1 mM) was used. An important advantage of this type of cultivation is that the metabolic shift from acidogenesis to solventogenesis and *vice versa* can simply be achieved by the adjustment of the external pH value above pH 5 (acidogenesis) or below pH 5 (solventogenesis).

It became apparent that the metabolic shift is associated with the synthesis of stress response proteins, e.g. GroEL, GroES, Hsp18, GrpE, HtpG, and DnaK (see also Section 3.2). Accumulation of stress proteins is not surprising as during the metabolic shift in parallel different stresses arise. Acidic stress is provoked by a decreasing pH and in consequence the concentration of undissociated acids increases. Furthermore, solvent formation itself caused stress for the cells. However, detection of stress proteins during the metabolic shift depended on the culture conditions. In cells of batch cultures an accumulation of stress proteins was clearly found whereas in chemostat cultures an increase of stress proteins was only transiently visible in the early phase of solventogenesis. Janssen *et al.* (2010)[16] revealed similar relative amounts of stress proteins under both metabolic conditions. This might be due to adaptation processes of exponentially growing cells in a chemostat culture.

A direct comparison of protein amounts in acidogenic cells and solventogenic cells using batch or continuous cultivation is presented in Table 3.3.

It seems logical that proteins directly involved in solvent formation were drastically increased or even only found after the metabolic shift. Prominent examples are the gene products of the *sol* operon, aldehyde dehydrogenase E (AdhE1) and the subunits of butyrate-acetoacetate CoA-transferase (CtfA, CtfB)[16,17], or acetoacetate decarboxylase (Adc)[15,18]. According to Janssen *et al.* (2010) NADH-dependent butanol dehydrogenase (BdhB) was also induced on the protein level in solventogenic cells[16]. Interestingly, most of the proteins typically related to the formation of acids largely remained unaffected in batch cultures as well as in chemostat cultures (Table 3.3). However, some of the glycolytic and acid formation enzymes showed conflicting results. Acetate kinase (Ack), butyryl-CoA dehydrogenase (Bcd), triosephosphate isomerase (Tpi), and pyruvate kinase (PykA) have been described to be downregulated in solventogenic cells of batch cultures[17,18], but showed unaltered amounts in solventogenic cells of steady-state chemostat cultures[16]. Notably, in solventogenic chemostat cells, the electron transfer flavoprotein beta-subunit (EtfB) was found in lower concentrations[16] and glyceraldehyde 3-phosphate dehydrogenase (GapC) was found in higher concentrations[15]. A comparative overview of described protein abundances is given in Table 3.3.

Different amounts with respect to the culture conditions used showed proteins involved in serine metabolism (D-3-phosphoglycerate

Table 3.3: Comparison of protein amounts in *C. acetobutylicum* cells of batch and continuous chemostat cultures before and after the metabolic switch from acidogenesis to solventogenesis.

Solventogenesis versus acidogenesis			**Proteome data of batch cultures**		**Proteome data of chemostat cultures**	
ORF[a]	Gene	Protein function	Sullivan[b]	Mao[c]	Schaffer[d]	Janssen[e]
Formation of solvents						
cap0162	*adhE1*	Aldehyde alcohol dehydrogenase	n.d.	↑	n.d.	↑
cap0163	*ctfA*	CoA-transferase A	n.d.	↑	n.d.	↑
cap0164	*ctfB*	CoA-transferase B	n.d.	↑	n.d.	↑
cap0165	*adc*	Acetoacetate decarboxylase	↑	↑	↑	↔
cac3298	*bdhB*	Butanol dehydrogenase	n.d.	↔	n.d.	↑
Formation of acids						
cac1742	*pta*	Phosphate acetyltransferase	n.d.	↔	n.d.	↔
cac1743	*ack*	Acetate kinase	n.d.	↓	n.d.	↔
cac2873	*thlA*	Acetyl-CoA acetyltransferase	n.d.	↔	n.d.	↔
cac2708	*hbd*	3-Hydroxybutyryl-CoA dehydrogenase	n.d.	↔	n.d.	↔
cac2709	*etfA*	Electron transfer flavoprotein	n.d.	↔	n.d.	↔
cac2710	*etfB*	Electron transfer flavoprotein	n.d.	↔	n.d.	↓
cac2711	*bcd*	Butyryl-CoA dehydrogenase	↓	↔	n.d.	↔
cac2712	*crt*	Enoyl-CoA hydratase	n.d.	↔	n.d.	↔
cac3075	*buk*	Butyrate kinase	n.d.	↔	n.d.	↔
cac3076	*ptb*	Phosphate butyryltransferase	n.d.	↑	n.d.	n.d.
Glycolysis						
cac2680	*pgi*	Glucose-6-phosphate isomerase	↓	↑	n.d.	↔
cac0517	*pfkA*	6-phosphofructokinase	n.d.	↔	n.d.	↔
cac0827	*fba*	Fructose-bisphosphatealdolase	↔	↔	n.d.	↔

Table 3.3: (*Continued*)						
cac0709	*gapC*	Glyceraldehyde 3-phosphate dehydrogenase	↔	↔	↑	n.d.
cac0710	*pgk*	Phosphoglycerate kinase	n.d.	↔	n.d.	↔
cac0711	*tpi*	Triosephosphate isomerase	↓	↔	n.d.	↔
cac0712	*pgm*	Phosphoglyceromutase	n.d.	↔	n.d.	↔
cac0713	*eno*	Phosphopyruvatehydratase	n.d.	↔	n.d.	↔
cac0518	*pykA*	Pyruvate kinase	n.d.	↓	n.d.	↔
Stress proteins						
cac1281	*grpE*	Molecular chaperone	n.d.	↑	n.d.	↔
cac1282	*dnaK*	Molecular chaperone	n.d.	↔	n.d.	↔
cac1283	*dnaJ*	Molecular chaperone	n.d.	↔	n.d.	n.d.
cac3315	*htpG*	Heat shock protein	n.d.	↑	n.d.	↔
cac3714	*hsp18*	Molecular chaperone	↑	↑	↑	↓*
cac2703	*groEL*	ChaperoninGroEL	↑	↑	n.d.	↔
cac2704	*groES*	Co-chaperonin	↑	↑	n.d.	↔
Selected proteins with different abundances						
cap0036		Uncharacterized protein	n.d.	n.d.	n.d.	↓
cap0037		Uncharacterized protein	n.d.	n.d.	n.d.	↓
cac0017	*serS2*	Seryl-tRNA-Synthetase	n.d.	↓	↑	n.d.
cac0014	*serC*	Aminotransferase	n.d.	↓	↑	n.d.
cac0015	*serA*	D-3-Phosphoglycerate dehydrogenase	n.d.	↓	↑	n.d.
cac0021	*serS1*	Seryl-tRNA-Synthetase	n.d.	↔	n.d.	↑
cac2584		Protein containing ChW-repeats	↑	↔	n.d.	↓
cac2203	*flaC*	Flagellin	n.d.	↓	n.d.	↑

[a]: Open reading frame[31].
[b]: Based on Sullivan and Bennett, 2006[18].
[c]: Based on Mao *et al.*, 2010[17].
[d]: Based on Schaffer *et al.*, 2010 ('onset of solventogenesis': cell sample was taken 6 h after the drop of the pH in the chemostat culture of *C. acetobutylicum* when steady-state growth conditions were not yet achieved)[15].
[e]: Based on Janssen *et al.*, 2010 (steady state)[16].
n.d.: Protein was not documented in the respective study.
↑: Significant higher protein abundance under conditions of solvent formation.
↔: No significant changes in protein abundance.
↓: Significant higher protein abundances under conditions of acid formation.
*: Identified protein as part of a spot containing more than one protein.

dehydrogenase, SerA; aminotransferase, SerC, and seryl-tRNA-synthetases, SerS1 and SerS2) and motility (flagellin, FlaC). After the induction of solvent production they all seemed to be upregulated in cells of chemostat cultures but downregulated in cells of batch cultures (Table 3.3). SerS1 was identified in two spots indicating different p*I* values (Table 3.4). In the case of FlaC, the apparent p*I* value and the apparent molecular weight differed significantly from the calculated values. A higher molecular weight might be explained by posttranslational high glycosylation of the native protein[43]. In *C. difficile*, glycosylated flagellins might be responsible for stable subunit–subunit interactions within the flagellar filament and efficient secretion of the flagellin monomer through the basal body apparatus[44]. Thus, it is conceivable that in solventogenic steady-state cells of chemostat cultures the export glycosylated FlaC protein is reduced or even blocked resulting in protein accumulation in the cytosol.

Janssen *et al.* (2010)[16] showed an extracellular neutral metallo-protease (CA_C2517) and a processive endoglucanase (CA_C0911), normally involved in cellulose degradation, to be upregulated in steady-state solventogenic cells (Table 3.4). Interestingly, a growth phase-dependent induction of these proteins has not been shown before. The reason for the significant upregulation of these specific enzymes in solventogenic cells is unknown, especially in an organism that is unable to utilize cellulose[45], although its genome comprises several genes encoding structural and enzymatic components of the cellulosome[46-48].

Striking results were obtained with respect to the two uncharacterized proteins CAP0036 and CAP0037, which are encoded by the megaplasmid pSOL1. Both proteins were found in considerable amounts only in acidogenic cells of chemostat cultures[16] and during the onset of solventogenesis[15]. Comparing phosphate-limited steady-state growth at pH 5.7 and 4.5, CAP0036 and CAP0037 revealed the strongest alterations in their amounts due to their disappearance in solventogenic cells. Both proteins were present in multiple spots, up to five in the case of CAP0037 and at least four in the case of CAP0036 (Table 3.4).

In each case, their horizontal spot distributions indicated three different isoelectric points. Additionally, CAP0037 revealed two

Table 3.4: Abundance of selected proteins and their characteristics in exponentially growing *C. acetobutylicum* cells in a chemostat culture under steady-state conditions of acidogenesis and solventogenesis[16] (modified).

ORF[a]	Protein name	Acido-genesis (steady state)	Solvento-genesis (steady state)	MW[b] (calc.)[c]	MW (app.)[d]	p*I*[e] (calc.)	p*I* (app.)
cac0911	Process. endoglu-canase	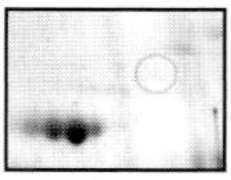	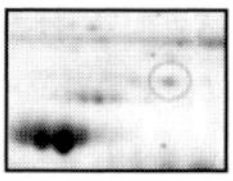	80.7	~80	5.12	~5.1
cac2517	Extracell. metallo-protease	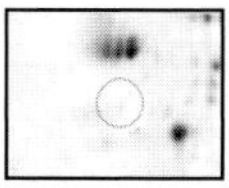	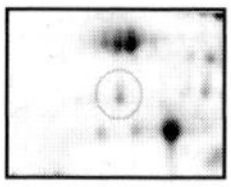	60.4	~43	6.38	~4.6
cac0021 (*serS1*)	Seryl-tRNA synthetase	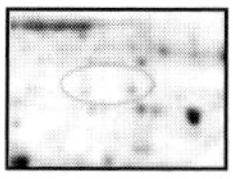	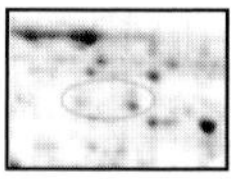	48.4	~49 ~48	5.75	~5.6 – ~5.8
cac2203 (*flaC*)	Flagellin	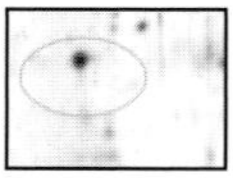	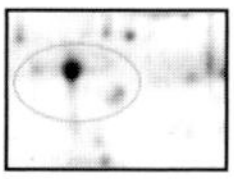	29.5	~40 ~39	5.78	~4.7 – ~4.9
cap0037	Uncharact. protein	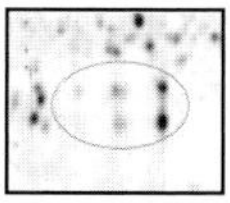	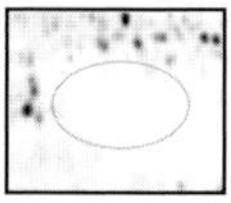	24.5	~25 ~22	5.58	~5.6 – ~6.0
cap0036	Uncharact. protein	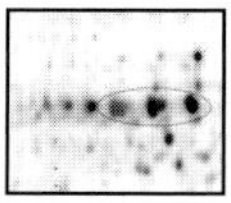	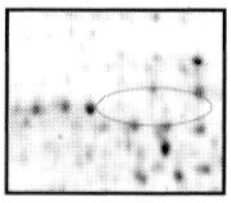	25.8	~30	5.53	~5.6 – ~6.0

[a]: Open reading frame[31].
[b]: Molecular weight (kDa) .
[c]: Calculated.
[d]: Apparent.
[e]: Isolelectric point.

different apparent molecular masses. It was speculated that this behaviour might indicate posttranslational modifications, such as phosphorylation, acetylation, glycosylation, or methylation, which have been reported in *C. acetobutylicum*[18,43,49] and other prokaryotic organisms[29,50].

CAP0036 and CAP0037 were annotated as orthologs of YgaT and YgaS of *Bacillus subtilis*[31]. Interestingly, they seemed restricted to some *Bacillus* and *Clostridium* species (Fig. 3.6). So far, only the nonsolvent *C. difficile* revealed a complete orthologous operon structure. In the ethanol-producing strain *C. phytofermentans* only a single open reading frame matching the *cap0036* gene and in the pathogenic nonsolvent *C. botulinum* only a gene orthologous to *cap0037* gene can be postulated[51].

Other orthologs might exist in *Bacilli* like *B. subtilis* (*ygaS*, formerly *yhbD*, Fig. 3.6), *B. thuringiensis*, *B. cereus*, and *B. licheniformis*. Usually, genes orthologous to *cap0037* are accompanied by genes resembling *cap0036* (in *B. subtilis ygaT*, formerly *yhbE*) in the same gene order (exception *C. difficile*: changed gene positions).

In contrast to *C. acetobutylicum*, other bacteria showed one (*C. difficile*, *B. cereus*) or up to six genes to be organized in the respective operon (*Bacillus anthracis*, *B. subtilis*, *B. thuringiensis*, *Bacillus weihenstephanensis*). In *B. subtilis*, orthologs are located in a three-gene operon (Fig. 3.6) (*yhbD–yhbE–yhbF*)[52].

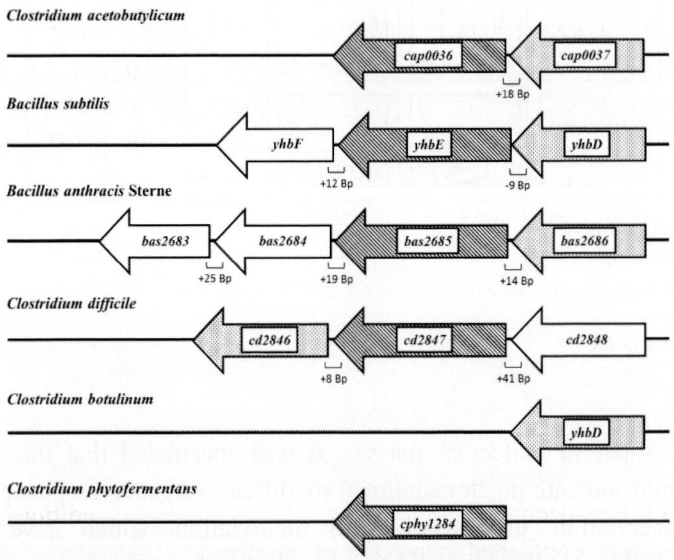

Fig. 3.6: Distribution of orthologs of *cap0037* and *cap0036* in other bacteria[51]. Orthologous genes are represented in the same design, *cap0037* (gray dotted), *cap0036* (black hatched) and further members of the same putative operon are white colored.

The biological functions of CA_P0036 and CA_P0037 in *C. acetobutylicum* and the reason for their high expression under conditions of acidogenic continuous steady-state growth remain unknown. The medium used may play a role in combination with phosphate limitation and pH. This could also explain why these proteins have not been noticed in earlier studies employing transient batch cultivation conditions.

3.5 Proteome data and transcriptome data, a comparison

One prerequisite for systems biology projects is the reliable integration of different sets of data. Today, transcriptome data play an important role. However, a high transcript level of a gene does not necessarily indicate an equivalent high amount of the respective protein in the cell and does not include any information about its activity[53].

So far, only a few comparative analyses of proteome and transcriptome data of *C. acetobutylicum* exist. In Tomas *et al.* (2004)[22] amounts of selected mRNAs in butanol-stressed cells were verified on the level of protein abundances by western-blot analysis (e. g., CtfA and CtfB). More often, proteome data gained in one laboratory[17,18] were compared with published transcriptome data[20,21,24] relying on the same culture conditions. The first study in which cell samples of the same chemostat culture were used for the generation of proteome data and transcriptome data was recently published by Janssen *et al.* (2010)[16]. Data for some proteins are summarized in Table 3.5.

A comparative overview of all proteome data and transcriptome data of *C. acetobutylicum* available in the literature is presented in Table 3.6. Although comparability of data to some extent is limited, results relying on both batch and on continuous cultivations using different media are combined. The focus is on relative changes from the viewpoint of solventogenic cells in relation to acidogenic cells.

A general understanding is that a high transcript level leads to a high amount of the respective protein. Independent of culture conditions, this hypothesis was confirmed by enzymes involved in the formation of solvents AdhE1, CtfA, and CtfB, which all are encoded by the *sol* operon. Similar behavior was shown for the enzyme catalyzing final step

Table 3.5: Comparison of protein abundances and transcript levels of selected proteins in acidogenic and solventogenic cells of a chemostat culture (steady-state growth conditions)[16].

ORF[a]	Protein name	pH 5.7 protein[b]	pH 4.5 protein[b]	pH 5.7 transcript[c]	pH 4.5 transcript[c]
Formation of solvents					
cap0162 (adhE1)	Aldehyde alcohol dehydrogenase		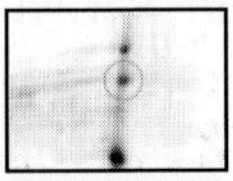	~100	~27000
cap0163 (ctfA)	CoA-transferase A	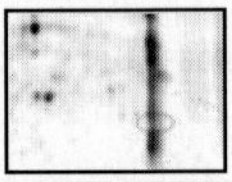	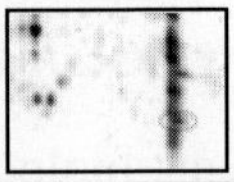	~100	~27000
cap0164 (ctfB)	CoA-transferase B	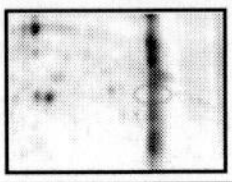	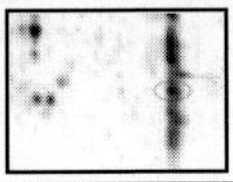	~100	~27000
cap0165 (adc)	Acetoacetate decarboxylase	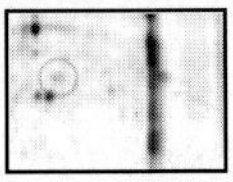	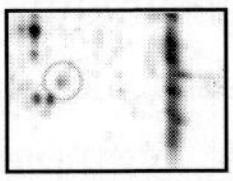	~2500	~5500
cac3298 (bdhB)	Butanol dehydrogenase B	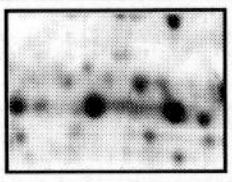	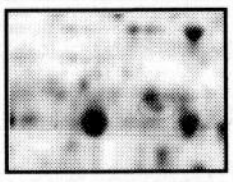	~2000	~2750
Formation of acids					
cac1742 (pta)	Phosphate acetyltransferase	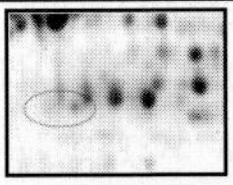	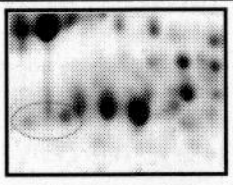	~1500	~1500
cac1743 (ack)	Acetate kinase	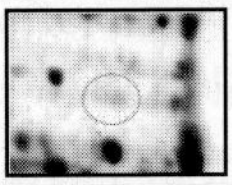	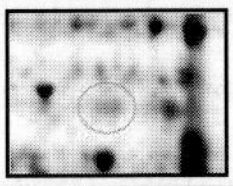	~1500	~1500
cac2873 (thlA)	Acetyl-CoA acetyltransferase		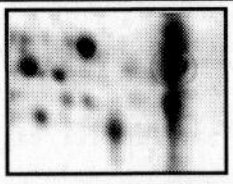	~16000	~3000
cac2708 (hbd)	3-hydroxybutyryl-CoA dehydrogenase	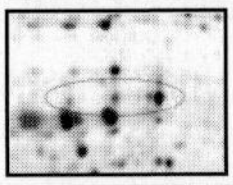	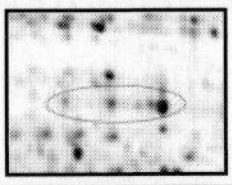	~5000	~1500
cac2709 (etfA)	Electron transfer flavoprotein	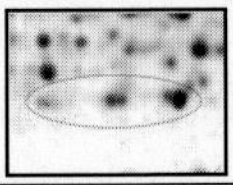	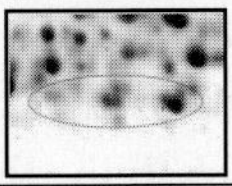	~15000	~4000
cac2710 (etfB)	Electron transfer flavoprotein			~12500	~2750

Table 3.5: (*Continued*)

cac2711 (*bcd*)	Butyryl-CoA dehydrogenase	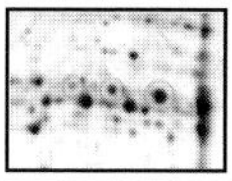	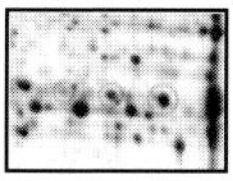	~14000	~3750
cac2712 (*crt*)	Enoyl-CoA hydratase	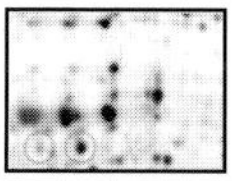	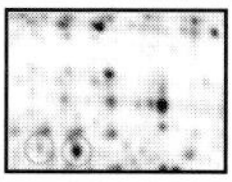	~12500	~3000
cac3075 (*buk*)	Butyrate kinase	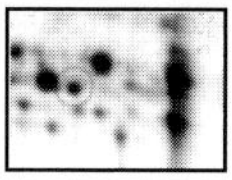		~5500	~1000

[a]: Open reading frame[31].
[b]: Protein amounts are represented by spot pictures.
[c]: mRNA transcript levels are expressed by "raw data" and the measured units of the responsible gene transcript under low scanning conditions of micro array evaluation[16].

of the formation of acetone (Adc). However, Janssen *et al.*, (2010)[16] documented only slightly elevated levels of mRNA molecules (ratio +2.5) and proteins (+1.25 fold) which did not match the significance criteria of ≥3 (transcriptome) and ≥2 (proteome).

More varying results have been reported with respect to proteins involved in acid formation. In solventogenic cells transcript levels of Hbd, EtfA, Bcd, Cr, t and Buk were clearly downregulated[16,24], but their protein amounts seemed not to be decreased, neither in batch[17] nor in continuous cultures[16] (Table 3.6). As an exception, the protein level of Bcd was decreased in solventogenic cells of a batch culture[18].

On the other hand, quantities of acetate kinase (Ack) mRNA remained constant[16,24,27] whereas its protein concentrations might depend on the culture conditions as they were found to be constant in chemostat cultures[16] but reduced in a batch culture[17].

Remarkably, a completely contrary behavior has been revealed for the phosphate butyryltransferase (Ptb) in cells of batch cultures. Its transcript level showed a significant downregulation in solventogenic cells[24], whereas its protein amount was found to be increased in the respective growth phase[17].

Most of the glycolytic enzymes listed in Table 3.6 revealed similar tendencies. Especially in cells of continuous cultures nearly no significant changes could be demonstrated on the level of transcription

Table 3.6: Comparison of the changes in protein amounts and mRNA concentrations of selected proteins of *C. acetobutylicum*.

Solventogenesis versus acidogenesis			Sullivan[b]	Mao[c]	Jones[d]	Schaffer[e]	Grimmler[f]	Janssen[g]	
ORF[a]	**Gene**	**Protein function**	**Protein**		**Transcript**	**Protein**	**Transcript**	**Protein**	**Transcr.**
Formation of solvents									
cap0162	*adhE1*	Aldehyde alcohol dehydrogenase	n.d.	↑	↑	n.d.	↑	↑	↑
cap0163	*ctfA*	CoA-transferaseA	n.d.	↑	↑	n.d.	↑	↑	↑
cap0164	*ctfB*	CoA-transferaseB	n.d.	↑	↑	n.d.	↑	↑	↑
cap0165	*adc*	Acetoacetate decarboxylase	↑	↑	↑	↑	↑	↔	↔
cac3298	*bdhB*	Butanol dehydrogenase B	n.d.	↔	↑	n.d.	↔	↑	↔
Formation of acids									
cac1742	*pta*	Phosphate acetyltransferase	n.d.	↔	↔	n.d.	↔	↔	↔
cac1743	*ack*	Acetate kinase	n.d.	↓	↔	n.d.	↔	↔	↔
cac2873	*thlA*	Acetyl-CoA acetyltransferase	n.d.	↔	↔	n.d.	↓	↔	↓
cac2708	*hbd*	3-hydroxybutyryl-CoA dehydrogenase	n.d.	↔	↓	n.d.	↓	↔	↔
cac2709	*etfA*	Electron transfer flavoprotein	n.d.	↔	↔	n.d.	↔	↔	↓
cac2710	*etfB*	Electron transfer flavoprotein	n.d.	↔	↔	n.d.	↔	↓	↔
cac2711	*bcd*	Butyryl-CoA dehydrogenase	↓	↔	↓	n.d.	↔	↔	↓
cac2712	*crt*	Enoyl-CoA hydratase	n.d.	↔	↓	n.d.	↓	↔	↓
cac3075	*buk*	Butyrate kinase	n.d.	↔	↓	n.d.	↔	↔	↓
cac3076	*ptb*	Phosphate butyryltransferase	n.d.	↑	↓	n.d.	↔	n.d.	↓

Table 3.6: (*Continued*)

cac2680	*pgi*	Glucose-6-phosphate isomerase	↓	↑	↔	n.d.	↔	↔	↔
cac0517	*pfkA*	6-phosphofructokinase	n.d.	↔	↔	n.d.	↔	↔	↔
cac0827	*fba*	Fructose-bisphosphate aldolase	↔	↔	↓	n.d.	↔	↔	↔
cac0709	*gapC*	Glyceraldehyde 3-phosphate dehydrogenase	↔	↔	↓	↑	↔	n.d.	↔
cac0710	*pgk*	Phosphoglycerate kinase	n.d.	↔	↔	n.d.	↔	↔	↔
cac0711	*tpi*	Triosephosphate isomerase	↓	↔	↔	n.d.	↔	↔	↔
cac0712	*pgm*	Phosphoglyceromutase	n.d.	↔	↔	n.d.	↔	↔	↔
cac0713	*eno*	Phosphopyruvate hydratase	n.d.	↔	↑	n.d.	↔	↔	↔
cac0518	*pykA*	Pyruvate kinase	n.d.	↓	↓	n.d.	↔	↔	↔
Stress proteins									
cac1281	*grpE*	Molecular chaperone	n.d.	↑	↑	n.d.	↑	↔	↔
cac1282	*dnaK*	Molecular chaperone	n.d.	↔	↑	n.d.	↑	↔	↔
cac1283	*dnaJ*	Molecular chaperone	n.d.	↔	↑	n.d.	↑	n.d.	↔
cac3315	*htpG*	Heat shock protein 90	n.d.	↑	↔	n.d.	↔	↔	↔
cac3714	*hsp18*	Molecular chaperone	↑	↑	↑	↑	↑	↓*	↔
cac2703	*groEL*	Chaperonin GroEL	↑	↑	↑	n.d.	↑	↔	↔
cac2704	*groES*	Co-chaperonin	↑	↑	↑	n.d.	↑	↔	↔

Table 3.6: (*Continued*)

Selected proteins with different abundance										
cap0036		Uncharacterized protein	n.d.	n.d.	↔	n.d.	↓	↓	↓
cap0037		Uncharacterized protein	n.d.	n.d.	↔	n.d.	↓	↓	↓
cac0017	*serS2*	Seryl-tRNA-Synthetase	n.d.	↓	n.d.	↑	↓	n.d.	↔
cac0014	*serC*	Aminotransferase	n.d.	↓	n.d.	↑	↔	n.d.	↔
cac0015	*serA*	D-3-Phosphoglycerate dehydrogenase	n.d.	↓	n.d.	↑	↔	n.d.	↔
cac0021	*serS1*	Seryl-tRNA-Synthetase	n.d.	↔	n.d.	n.d.	↔	↑	↔
cac2584		Protein containing ChW-repeats	↑	↔	n.d.	n.d.	↔	↓	↔
cac2203	*flaC*	Flagellin	n.d.	↓	↓	n.d.	↓	↑	↔

[a]: Open reading frame[31]; [b]: based on Sullivan and Bennett, 2006[18]; [c]: based on Mao *et al.*, 2010[17].
[d]: Based on Jones *et al.*, 2008[24]; [e]: based on Schaffer *et al.*, 2010 ('onset of solventogenesis'; refer to Table 3)[15].
[f]: Based on Grimmler *et al.*, 2011[27]; [g]: based on Janssen *et al.*, 2010 (steady state)[16].
n.d.: Protein/transcript was not documented.
↑: Significantly higher protein abundance/transcript level in solventogenesis.
↔: Protein or transcript was identified but no significant changes in protein abundance/transcript level between both growth phases occurred.
↓: Significantly higher protein abundance/transcript level in the acidogenesis.
*: Identified peptide was part of a spot which contained more than one protein.

and translation, respectively. Cells of batch cultures showed slightly different behavior. For example contrary amounts of glucose-6-phosphate isomerase (Pgi) were described[17,18]. Furthermore, in contrast to chemostat cells, mRNA and protein amounts of pyruvate kinase (PykA) were found to be downregulated after the solventogenic shift. It is noteworthy that, with respect to glycolytic enzymes, the study of Jones *et al*. (2008)[24] showed some differences in comparison with the results of the other groups.

As might be expected from the Section 3.2 ("Stress proteomes of *C. acetobutylicum*"), several stress response candidates (GrpE, Hsp18, GroEL, and GroES) showed both high transcript levels and also high protein amounts in reaction to the solventogenic switch. However, in continuous cultures this finding seemed to be restricted to the onset of solventogenesis indicating a transient upregulation[27]. This could explain why during steady-state conditions of growth no significant alterations were monitored[16]. It can be postulated that this behavior might be due to adaptation processes in the cells. Furthermore, in batch cultures much higher concentrations of solvents were reached (~250 mM, acetone plus butanol) in comparison to continuous cultures (~70 mM, acetone plus butanol). In consequence, higher stress conditions are caused keeping the stress response on an increased level.

The last part of Table 3.6 contains candidates with different appearances. Results dealing with CA_P0036, CA_P0037 are discussed above. With respect to the proteins involved in serine metabolism, opposite regulations in proteome data were revealed by Schaffer *et al.* (2002)[15] in comparison to Mao *et al.* (2010)[17].

A contrariwise regulation between protein and transcript levels was also shown for flagellin (FlaC) by Janssen *et al.* (2010)[16] (see above, Section 3.4; Table 3.4). Other than this, all studies revealed significant decreases of FlaC. This might be a good example demonstrating differences due to individual cultivation conditions.

3.6 Conclusions and outlook

Proteome analysis of *C. acetobutylicum* delivered valuable data for the understanding of the metabolism of this bacterium. This method represents one of the oldest "-omic" techniques, but currently more and more "-omic" technologies, such as transcriptomics, are becoming important. The integration of multiple "-omic" approaches for one experiment is important, because any single "-omic" approach probably is insufficient to explain and characterize the complexity of a regulatory network in a cell[54].

With respect to systems biology projects it is necessary that the data sets generated by the different "omics" technologies are comparable. Therefore, the development and application of reliable standard operating procedures is absolutely mandatory. This applies not only to the composition of culture media and cultivation procedures, but also to sample preparation. Furthermore, the use of identical cell samples for different "-omic" analyses would be ideal.

Moreover, data of the individual "-omic" techniques have to be interpreted carefully and limitations of the method must be taken into account. Proteome analysis is time consuming and covers only a part of the deduced gene products of the genome. Minor abundant proteins may be overlooked. On the other hand, as demonstrated in this chapter, a high transcript level does not necessarily reflect an equivalent high amount of the respective protein. Furthermore, posttranslational regulatory mechanisms such as protein modifications cannot be monitored by transcriptomics but can be at least indicated by proteomics.

Within the different "-omic" technologies, proteome analysis plays an important role because of its ability to image protein patterns directly. Thus, proteome analysis will be helpful in supporting systems biology approaches in the future, although perhaps not as its key technology. Its importance will depend on further technological and methodological developments. One objective might be the visualization of protein complexes or the resolution of microcompartments of the bacterial cell[55].

It should be mentioned that proteome studies of *C. acetobutylicum* have mostly covered the protein fraction of the cytosol so far. The analyses of the secretome and membrane proteins, cell wall, or cell wall-associated proteins[19] are still in their infancy and have to be investigated in more detail in the future.

Acknowledgement

We thank the German Federal Ministry of Education and Research (BMBF) for financial support (project COSMIC-SysMO, 0313981D).

References

1. De Keersmaecker S. C. J. *et al.*, *Mol. Microbiol.* 62 (2006).
2. Joyce A. R. and Palsson B. Ø., *Nat. Rev. Mol. Cell. Biol.* 7 (2006).
3. Millat T. *et al.*, *Methods Mol. Biol.* 696 (2011).
4. Dürre P., *Biotechnol. J.* 2 (2007).
5. Dürre P., *Ann. NY Acad. Sci.* 1125 (2008).
6. Lee S. Y. *et al.*, *Biotechnol. Bioeng.* 101 (2008).
7. Heap J. T. *et al.*, *J. Microbiol. Methods* 70 (2007).
8. Heap J. T. *et al.*, *J. Microbiol. Methods* 70 (2010).
9. O`Farrell P. H., *J. Biol. Chem.* 250 (1975).
10. Bahl H., Ph.D. Thesis, University of Göttingen, Germany (1983)
11. Terracciano J. S. *et al.*, *Appl. Environ. Microbiol.* 54 (1988).
12. Pich A. *et al.*, *Appl. Microbiol. Biotechnol.* 33 (1990).
13. Kawasaki S. *et al.*, *FEBS Letters* 571 (2004).
14. May A. *et al.*, *FEMS Microbiol. Let.* 238 (2004).
15. Schaffer S. *et al.*, *Electrophoresis* 23 (2002).
16. Janssen H. *et al.*, *Appl. Microbiol. Biotechnol.* 87 (2010).
17. Mao S. *et al.*, *J. Proteome Res.* 9 (2010).
18. Sullivan L. and Bennett G. N., *J. Ind. Microbiol. Biotechnol.* 33 (2006).
19. Mao S. *et al.*, *Mol. Biosyst.* 7 (2011).
20. Harris L. M. *et al.*, *J. Bacteriol.*184 (2002).
21. Tomas C.A. *et al.*, *J. Bacteriol.* 185 (2003).
22. Tomas C. A. *et al.*, *J. Bacteriol.*186 (2004).
23. Alsaker K. V. and Papoutsakis E. T., *J. Bacteriol.* 187 (2005).
24. Jones S.W. *et al.*, *Genome Biol.* 7 (2008).
25. Hillmann F. *et al.*, *J. Bacteriol.* 191 (2009).
26. Grimmler C. *et al.*, *J. Biotechnol.* 150 (2010).
27. Grimmler C. *et al.*, *J. Mol. Microbiol. Biotechnol.* 20 (2011).
28. Mann M. and Jensen O. N., *Nat. Biotechnol.* 21 (2003).
29. Rosen R. and Ron E. Z., *Mass. Spec. Rev.* 21 (2002).
30. Haus S. *et al.*, *BMC Systems Biology* 5:10 (2011).
31. Nölling J. *et al.*, *J. Bacteriol.* 183 (2010).
32. De Maré F. *et al.*, *Nat. Struct. Biol.* 3 (1996).
33. Hillmann F. *et al.*, *Mol. Microbiol.* 68 (2008).
34. Riebe O. *et al.*, *Microbiology* 155 (2009).

35. Fiedler T. *et al.*, *J. Bacteriol.* 190 (2008).
36. Schwarz K. *et al.*, *J. Microbiol. Meth.* 68 (2007).
37. Schwarz K., Ph.D. Thesis, University of Rostock, Germany (2007).
38. Desvaux M. *et al.*, *Biochim. Biophys. Acta.* 1745 (2005).
39. Fiedler T., Ph.D. Thesis, University of Rostock, Germany (2006)
40. Tatusov R. L. *et al.*, *Nucleic Acids Res.* 28 (2000).
41. Hiller K. *et al.*, *Bioinformatics* 22 (2006).
42. Bahl H. *et al.*, *Eur. J. Appl. Microbiol. Biotechnol.* 15 (1982).
43. Lyristis *et al.*, *Anaerobe* 6 (2000).
44. Twine S. M. *et al.*, *J. Bacteriol.* 191 (2009).
45. Lee S. F. *et al.*, *Appl. Environ. Microbiol.* 50 (1985).
46. Sabathé F. *et al.*, *FEMS Microbiol. Lett.* 217 (2002).
47. López-Contreras A. M. *et al.*, *Appl. Environ. Microbiol.* 69 (2003).
48. López-Contreras A. M. *et al.*, *Appl. Environ. Microbiol.* 70 (2004).
49. Balodimos I. A. *et al.*, *Appl. Environ. Microbiol.* 56 (1990).
50. Rosen R. *et al.*, *Proteomics* 4, 3068–3077 (2004).
51. Karp P. D. *et al.*, *Nucleic Acids Res.* 33 (2005).
52. Floréz L. A. *et al.*, *SubtiWiki Database (Oxford)* (2009).
53. Zhang W. *et al.*, *Microbiology* 156 (2010).
54. Gygi S.P. *et al.*, *Mol. Cell Biol.* 19 (1999).
55. Yeates T. O. *et al.*, *Annu. Rev. Biophys.* 39 (2010).

CHAPTER 4

COMPARATIVE GENOMIC ANALYSIS OF THE GENERAL STRESS RESPONSE IN *CLOSTRIDIUM ACETOBUTYLICUM* ATCC 824 AND *CLOSTRIDIUM BEIJERINCKII* NCIMB 8052

K. M. SCHWARZ

Centre for Biomolecular Sciences,
BBSRC Sustainable Bioenergy Centre University Park
University of Nottingham, Nottingham, NG7 2RD, UK

S. W. M. KENGEN

Laboratory of Microbiology Wageningen University
Dreijenplein 10, 6703HB, Wageningen, NL

Among the numerous solventogenic clostridial strains that have been isolated, two species have emerged as model organisms for studying butanol production, *viz. Clostridium acetobutylicum* ATCC 824 and *Clostridium beijerinckii* NCIMB 8052. The metabolic switch from acid production to solvent production which is typical for these organisms is coupled to spore formation but also general stress response. By the availability of the genome sequences of both model organisms, we were able to compare the genetic organization of their stress response. High similarities existed regarding the presence of genes encoding heat shock proteins, their classification into established *Bacillus subtilis hsp* gene categories as well as their organization within operons and their regulatory structures. *C. acetobutylicum* and *C. beijerinckii* possess class I and III *hsp* genes, with class III *hsp* genes including members of the *clpC* operon. Both organisms also possess genes for potential class IV *hsp* genes, however, they do not appear to contain class II type *hsp* genes. Similarities and differences in the organization of the genes and gene clusters are discussed.

4.1 Introduction

The heat shock response is one of the best-studied stress response networks. Its strong evolutionary preservation in prokaryotes and eukaryotes, the ubiquity and sequence conservation of the involved major heat shock proteins (HSPs), and their induction by a variety of environmental stress conditions have led to the assumption that the heat shock response represents rather a general stress response[1]. Due to their protein modifying character, the HSPs are mainly representatives of the Category O (posttranslational modification, protein turnover, chaperone functions) of the clusters of orthologous groups (COG)[2,3]. Functionally they can be separated into two groups: molecular chaperones and proteases. Whilst molecular chaperones like GroES (HSP10), GroEL (HSP60), DnaK, and DnaJ (both HSP70) play an essential role in the synthesis, folding, and transport of proteins in stressed and non-stressed cells, proteases such as Lon- and Clp-family proteases (HSP100 family) take action in the degradation of misfolded proteins[4]. In the Gram-positive bacterium *Bacillus subtilis*, genes coding for the general stress response (often referred to as heat shock protein [*hsp*] genes) have been grouped into four classes based on their mechanism of regulation. Thus, class I *hsp*-genes are represented by genes whose expression depends on the housekeeping sigma factor σ^A, a highly conserved *cis*-acting 9-bp inverted repeat called CIRCE element (controlling inverted repeat of chaperone expression, consensus sequence: TTAGCACTC-N9-GAGTGCTAA) and the heat-inducible transcription repressor HrcA. Class I genes are generally genes coding for the molecular chaperones DnaK, DnaJ, GrpE, GroES, and GroEL[5]. Heat, salt, ethanol, or starvation-inducible *hsp*-genes which are positively regulated by the alternative sigma factor σ^B are designated as class II genes. Class III gene expression is controlled by the regulator protein CtsR (class three stress gene regulator) and a 7-bp tandem *cis*-regulatory element (consensus sequence: A/GGTCAAA-NAN-A/GGTCAAA[6]) in the promoter region of the target genes. Representatives are the heat-inducible members of the Clp-family of proteases required for cellular processes like motility, sporulation, and competence[6,7]. Other genes, known to be heat inducible and of general cellular function but with an unknown or different mechanism of regulation, are summarized as class IV *hsp* genes[5,8,9]), although more detailed classifications for *B. subtilis* have been already suggested[10].

In the past, the heat shock response of *C. acetobutylicum* was investigated to some extent and identified *hsp* genes were grouped using the above discussed classification of its close relative *B. subtilis*[9,11–18]. In contrast, almost nothing is known about the *C. beijerinckii hsp* genes. However, by the availability of the genome sequence of both clostridia, a detailed comparison on gene level became possible. The purpose of this chapter is to give an overview of the general stress response proteins and their encoding genes in *C. beijerinckii* based on the current knowledge in *C. acetobutylicum*. Searching the *C. beijerinckii* genome for known *C. acetobutylicum hsp* genes, orthologs for all investigated genes were found[3,19] (Table 4.1). These orthologs belong to one or another COG O subclass and can be, like in *C. acetobutylicum*, allocated to one of the *hsp* gene classes. This classification will be discussed in more detail in the following paragraphs.

4.2 Class I heat shock proteins

A search in the *C. beijerinckii* genome for genes encoding the known *C. acetobutylicum* class I *hsp* genes (*dnaK*, *dnaJ*, *grpE*, *groES*, *groEL*, and *hsp90*) resulted in significant matches displaying the same gene organization and a high sequence identity and length similarity of the deduced amino acid sequences (Fig. 4.1A). In both organisms, the *dnaK* gene region encodes a heat-inducible transcriptional repressor HrcA (CA_C1280, CBEI_0828) and the molecular chaperones GrpE (CA_C1281, CBEI_0829), DnaK (CA_C1282, CBEI_0830), and DnaJ (CA_C1283, CBEI_0831). Both gene regions are located on the leading strand, in *C. acetobutylicum* in the middle of the 3.94-Mbp genome[20] and in *C. beijerinckii* in the beginning of the 6-Mbp genome[21,22] (http://genome.ornl.gov/microbial/cbei/). Within the *C. acetobutylicum dnaK* gene region several transcription start sites (TSS) with corresponding promoter sequences as well as a rho-independent terminator[23-25] downstream of the *dnaK* gene have been identified. Thereby, especially the heat-inducible promoter located directly in front of *hrcA* displayed a high homology to the consensus promoter sequence recognized by the vegetative sigma factor σ^A of Gram-positive bacteria and is in close proximity to a CIRCE element[15,26] (Fig. 4.1A). In *C. beijerinckii* at least one σ^A housekeeping promoter in close proximity to a CIRCE element

Table 4.1: COGO subclass representatives in *C. acetobutylicum* (*Cac*) and *C. beijeinckii* (*Cbei*).

COG ID	*Cac* locus[a]		name[b]	class[c]	*Cbei* locus		name	% identitiy [d]
0071	CA_C3714	HSP18	small heat shock protein	III[e]	Cbei_4123	HSP20	heat shock protein Hsp20	44
					Cbei_5030		hypothetical protein	26
0234	CA_C2704	GroES	Co-chaperonin GroES (HSP10)	I	Cbei_0328	GroES	co-chaperonin GroES	72
0265	CA_C0463		Serine protease Do (heat-shock protein)		Cbei_0379		2-alkenal reductase	36
	CA_C0500		C-terminal PDZ domain-containing protein		Cbei_4865		C-terminal PDZ domain-containing protein	37
	CA_C2433	HtrA	HtrA-like serine protease	IV	Cbei_0379		2-alkenal reductase	42
	CA_C3218		trypsin-like serine protease		Cbei_0379		2-alkenal reductase	32
0326	CA_C3315	HSP90	Molecular chaperone, HSP90 family	I	Cbei_4160		heat shock protein 90	33
					Cbei_1614		hypothetical protein	29
0443	CA_C0472	DanK	HSP70/DnaK family protein		Cbei_0830	DanK	molecular chaperone DnaK	42
	CA_C0473	DanK	heat shock protein DnaK		Cbei_0830		molecular chaperone DnaK	35
					Cbei_4397		molecular chaperone-like protein	nsm*
	CA_C1282	DnaK	molecular chaperone DnaK	I	Cbei_0830	DanK	molecular chaperone DnaK	79
0459	CA_C2703	GroEL	chaperonin GroEL	I	Cbei_0329	GroEL	chaperonin GroEL	82
0465	CA_C0602	FtsH	ATP-dependent zinc metallopeptidase FtsH		Cbei_4783		ATP-dependent metalloprotease FtsH	64
					Cbei_3037		ATP-dependent metalloprotease FtsH, extracellular	61
	CA_C0955		ATP-dependent Zn protease, extracellular		Cbei_2530		ATP-dependent metalloprotease FtsH, extracellular	56
	CA_C3202	FtsH	ATP-dependent Zn protease, FTSH, extracellular	IV	Cbei_0100		ATP-dependent metalloprotease FtsH, extracellular	68
0466	CA_C0456	LonA	ATP-dependent protease (lonA)	IV	Cbei_1254		ATP-dependent protease La	64
	CA_C2637	LonA	ATP-dependent Lon protease	IV	Cbei_1330		ATP-dependent protease La	61
0484	CA_C0648		molecular chaperone		Cbei_1414		heat shock protein DnaJ domain-containing protein	47
					Cbei_3106		heat shock protein DnaJ domain-containing protein	nms*

	CA_C1283	DnaJ	DnaJ family molecular chaperone	I	Cbei_0831		chaperone protein DnaJ	67
					Cbei_4404		heat shock protein DnaJ domain-containing protein	58
0542	CA_C0904		ABC transporter ATPase		Cbei_0123		ATPase	51
	CA_C0959	ClpB	ABC transporter ATPase		Cbei_0645		ATPase	70
	CA_C1824	ClpA	ATP-dependent Clp proteinase	IV	Cbei_4152		ATPase	54
	CA_C3189	ClpC	ABC transporter ATPase	III	Cbei_0123		ATPase	64
0576	CA_C0471		GrpE protein HSP-70 cofactor		-		-	-
	CA_C1281	GrpE	molecular chaperone GrpE	I	Cbei_0829		GrpE protein	68
0740	CA_C1811		periplasmic serine protease		Cbei_1212		peptidase S14, ClpP	65
	CA_C1893		ClpP family serine protease, possible phage related		Cbei_4002		endopeptidase Clp	26
					Cbei_1327		ATP-dependent protease, ATP-binding subunit	26
	CA_C2640	ClpP	ATP-dependent Clp protease, proteolytic subunit	IV	Cbei_1327		ATP-dependent protease, ATP-binding subunit	75
					Cbei_4002		endopeptidase Clp	71
1067	CA_C2135		ATP-dependent serine protease		Cbei_0810		ATP-dependent protease	43
	CA_C2638	LonB	Lon-like ATP-dependent protease		Cbei_1329		sporulation protease LonB	56
	CA_C3716	LonB	Lon-like ATP-dependent protease		Cbei_5080		sporulation protease LonC	64
1219	CA_C2639	ClpX	ATP-dependent protease, ATP-binding subunit	IV	Cbei_1328	ClpX	ATP-dependent protease, ATP-binding subunit	76

a systematic locus tag
b annotated name
c classification of *C. acetobutylicum* HSP according to *B. subtilis hsp* gene categorization[9]
d amino acid percent identity[19] [algorithm: blastp, updated on 26.11.2011]
e see text for further explanations

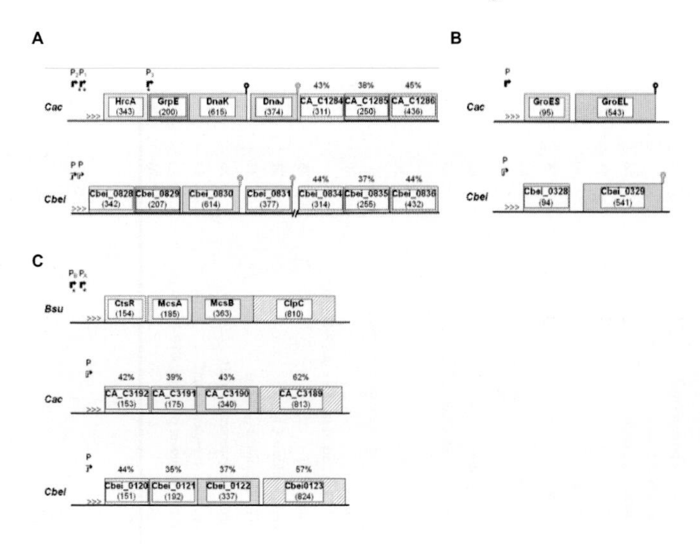

Fig. 4.1: Comparison of the (A) *dnaK*, (B) *groESL*, and (C) *clpC* gene regions of *C. acetobutylicum* (*Cac*) and *C. beijerinckii* (*Cbei*). The same shading refers to similar genes. Length of the amino acid sequence is shown in brackets. Black bold arrows indicate confirmed promoter sequences, and black light-shaded arrows indicate theoretical promoter sequences. Arrow heads indicate the location of the CIRCE element and bold asterisks the location of proven transcription start sites (TSS). Bold- and light-shaded hairpins mark already described and theoretical transcription terminators, respectively. Amino acid sequence identities for (A) and (B) are given in Table 4.1, except for the last three encoded polypeptides of the *dnaK* gene region of *C. beijerinckii*. As the existence of a heptacistronic operon in *C. acetobutylicum* has not been confirmed yet, data values concern the sequence identity between *B. subtilis* and *C. beijerinckii*. As for both *C. acetobutylicum* and *C. beijerinckii* a *clpC* an operon is yet to be analyzed, (C) compares the putative *clpC* gene region of the two species with the *B. subtilis* (*Bsu*) *clpC*-operon. In this case, per cent values state the identity to the deduced *B. subtilis* polypeptides. Recently, CBEI_0124 has been predicted to be part of the *clpC* operon[22]. All amino acids similarity/identity searches were carried out on 26.11.2011.

upstream of *hrcA* (CBEI_0828) could be identified (Fig. 4.2A). This resembles the situation in *C. acetobutylicum* and classifies the members of the *C. beijerinckii dnaK* gene region as class I *hsp* genes[5,9]. A nucleotide sequence holding common features of a prokaryotic rho-independent transcription terminator with a calculated minimum free energy of $\Delta G = -82.5$ kJ/mol has been identified between the *dnaK* and

A

Ca_C1280 (*hrcA*)* -AAG**TTGACA**A-[15]-A**TATTTT**A-[11]-ATTAGCACTCAAAGAGAGTGAGTGCTAATAA- 12

CA_C2704 (*groES*)* -ATGT**TTGCTA**A-[15]-T**TATTAT**A-[8]-GTTAGCACTCAAGATTAACGAGTGCTAACAA- 76

CA_C3315 (*hsp90*)* -GTA**TTGACA**A-[14]-C**TAAAAT**A-[2]-ATTAGCACTCGTTGATATAGAGTGCTAACAA- 86

Cbei_0828 (*hrcA*)* -GAA**TTGACA**A-[15]-A**TATATT**A-[11]-ATTAGCACTCGAGGTTAATGAGTGCTAACAA- 14

Cbei_0328 (*groES*)* -ATGT**TTGAAT**T-[18]-T**TATAAT**T-[2]-GTTAGCACTCGACAATGATGAGTGCTAACAG- 45

B

CA_C3192 (*ctsR*) -ATAGTA**TTGAAA**A-[15]-A**TATAAT**A-[1]-AGTCAAAGAAAGTCAAAGTA- 25

Cbei_0120 (*ctsR*) -TGATGC**TTCAAA**A-[15]-G**TATAAT**A-[6]-AGTCAAAGTAAGTCAAAAAC- 34

Fig. 4.2: Upstream promoter region of (A) class I and (B) class III *hsp* genes. Shown are the upstream promoter regions of *C. acetobutylicum* (Cac) and *C. beijerinckii* (Cbei) genes possessing a (A) a CIRCE element or (B) a CtsR binding site. The respective *cis* regulatory elements are underlined. Sequences resembling the promoter sequences of clostridia are marked in bold, whereby sequences to the left mark the –35 box and sequences to the right the –10. Numbers to the right indicate the distance from the start ATG before displaying the sequence. Asterisks next to the locus tag indicate the existence of usually weaker promoter sequences upstream of the indicated ones (for CA_C1280 see[16], for CA_C2704 the –10 box at 37 bp and the –35 box at 58 bp upstream of start ATG, for CA_C3315 the –10 box at 37 bp and the –35 box at 43 bp upstream of start ATG, for CBEI_0328 the –10 box at 125 bp upstream of start ATG and no obvious –35 box found, for CBEI_0828 the –10 box at 100 bp and the –35 box at 135 bp upstream of start ATG).

dnaJ gene of *C. acetobutylicum* (Fig. 4.2A, bold hairpin)[16,27]. A sequence with similar features could be identified 7-bp downstream of the stop TAA of the *C. beijerinckii dnaK* as well[23–25,27] (Fig. 4.1A, light-shaded hairpin). It comprises a 12-bp stem with one unpaired U and a 7-bp loop (AAACUAC). The U-rich sequence (UUUUCU) is with 33 bp rather far away from the stem loop. The calculated minimum free energy of ΔG = –98.32 kJ/mol[27] is comparable to that of *C. acetobutylicum* (see above). Beyond that, the *dnaK*-operon of *B. subtilis* has been described as a heptacistronic operon comprising in the following order the genes *hrcA* (BSU25490), *grpE* (BSU25480), *dnaK* (BSU25470), *dnaJ*

(BSU25460), *yqeT* (BSU25450), *yqeU* (BSU25440), and *rimO* (BSU25430). *yqeT, yqeU*, and *rimO* code for a potential ribosomal protein L11 methyltransferase, a 16S ribosomal RNA methyltransferase RsmE and a ribosomal protein S12 methylthiotransferase. Their genes are transcribed together with all other genes of the *dnaK*-operon as well as a sub-operon starting at an internal vegetative promoter upstream of *dnaJ*. They are not involved in the heat shock response and are not essential for growth under elevated temperatures[10,28]. In *C. acetobutylicum*[28] and *C. beijerinckii* genes coding for polypeptides with the highest identity to the deduced amino acid sequences of these *B. subtilis* genes are located downstream of *dnaJ*, in *C. acetobutylicum* directly and in *C. beijerinckii* approximately 3.5 kbp further downstream of *dnaJ* succeeding two embedded open reading frames (orf)[19] (Fig. 4.2A, light-shaded hairpin). This enables for *C. acetobutylicum* the possibility of a heptacistronic operon like in *B. subtilis*. Moreover, the potential rho-independent transcription terminator (16-bp stem including an U-G and an A-G mismatch and a 5-bp loop [UAUUU], followed in 15 bp by a U-rich sequence [UUAUAAUUUAUU]), has, with a calculated free energy of $\Delta G = -58.77$ kJ/mol[27], a lower folding stability than the one after the *dnaK* gene (CA_C1282). A potential read through has already been observed for the rho-independent terminator following *dnaK*[16], thus, making a read through at this position even more likely to happen. Nevertheless, experimental data on the function of this theoretical terminator as well as the existence of a heptacistronic *C. acetobutylicum dnaK* operon are not available yet. A potential terminator sequence 10 bp downstream of the *C.beijerinckii dnaJ* gene has a higher calculated free energy ($\Delta G = -85.7$ kJ/mol). This makes, in comparison to *C. acetobutylicum*, a transcriptional stop at this point more reasonable[23–25,27] and, therefore, militating in favor of a *C. beijerinckii* operon just comprised of *hrcA-grpE-dnaK-dnaJ*[22]. The terminator is characterized by a 23-bp stem comprising a 5-bp loop [UCAAU] and an U-rich sequence [UUUCUUUUUU] which follows directly.

The *groESL*-operon of *C. acetobutylicum* is bicistronic (CA_C2704-CA_C2703). Unlike the *dnaK* gene region of the organism it is located on the lagging strand at the end of the 3.94-Mbp genome. It possesses, like the *dnaK* gene region, a heat-inducible σ^A promoter, a CIRCE element, and a rho-independent terminator downstream of *groEL*[15,26] (Fig. 4.2B and 2A). In contrast, the *C. beijerinckii groESL*-operon is located on the leading strand in the beginning of the 6-Mbp genome. It is represented by the orfs with the locus tags CBEI_0328 (*groES*) and CBEI_0329 (*groEL*)[22] (Fig. 4.2B and 2A). The identified potential terminator sequence 35 bp downstream of *groEL* consists of a 19-bp stem structure with a 4-bp loop (UAUA) and is in a distance of 3 bp followed by a U-rich sequence [UUUUU]. The calculated free energy of $\Delta G = -100$ kJ/mol suggests, compared to other *C. acetobutylicum* terminators[16,29–31], a stop of the termination.

In *C. acetobutylicum*, the gene encoding HSP90 (*htpG*, CA_C3315) was due to its possession of a σ^A promoter and a CIRCE element (Fig. 4.2A), classified as class I *hsp*[17,18]. It is encoded on the lagging strand and surrounded by genes potentially encoding a nitroreductase family protein (CA_C3314) and a cardiolipin synthase (CA_C3316). The gene with the highest similarity in *C. beijerinckii* is located on the lagging strand and annotated as HSP90 (CBEI_4160). However, its deduced amino acid sequence shows only a sequence identity of 33% to the *C. acetobutylicum* HSP90 (CA_C3315)[19]. No CIRCE element was found in the promoter region of this gene classifying it like the *hsp90* (*htpG*) of *B. subtilis* as class IV *hsp*[9,32].

Carrying out a whole genome analysis of *C. acetobutylicum* and *C. beijerinckii* using the search algorithm provided by the virtual footprint analyzer[33] no further class I candidates were found in either organism.

4.3 Class II heat shock proteins

In contrast to *B. subtilis*, no class II *hsp* genes could be identified in *C. acetobutylicum* and *C. beijerinckii*. Both organisms lack the alternative sigma factor σ^B and other gene products of its encoding operon (*rsbRSTUVW-sigB-rcbX*)[11,12,17,19,34]. Furthermore, typical *B. subtilis* class II stress polypeptides, such as YflT, GspA, and BmrU, could not be identified in *C. acetobutylicum* and *C. beijerinckii*. Other genes coding for gene products with homology to known *B. subtilis* class II HSPs might, therefore, be subjected to another so far unknown regulation mechanism.

4.4 Class III heat shock proteins

In *C. acetobutylicum*, a sequence possessing high similarity to the *B. subtilis clpC* operon known to encode class III *hsp* genes has already been identified but is yet to be further described[17,19] (Fig. 4.1C). This region is located on the lagging strand and comprises the orfs with the locus tags CA_C3192, CA_C3191, CA_C3190, and CA_C3189. In analogy to the *clpC* operon of *B. subtilis*, the first *C. acetobutylicum* gene potentially encodes for the transcriptional repressor CtsR and the fourth for ClpC, a representative of the adenosine triphosphate- (ATP-) dependent Clp-family of proteases[9,35–38]. These proteases are often, as with Lon-family proteases, described as HSP100[39] and are involved in the regulation of many cellular processes, including heat shock response, protein transport, and degradation[40,41]. The gene product of the second gene of the *C. acetobutylicum clpC* gene region (CA_C3191) possesses, in spite of its annotation as hypothetical protein, the highest amino acid sequence identity (38%) to McsA (BSU00840) (Fig. 4.2C)[19], the second gene of the *B. subtilis clpC*-operon and activator of the CtsR modulator McsB. In addition, it possesses, like the *B. subtilis* McsA, zinc finger motifs (CXXC, from start methionine: 3–6, 29–32, 84–87, 102–105) which are thought to be involved in the interaction of McsA with McsB in *B. subtilis*. The polypeptide encoded by the third orf of the *C. acetobutylicum clpC* gene region (CA_C3190) shows the greatest similarity to the *B. subtilis* McsB (BSU00850) (Fig. 4.1C)[42]. Like its *B. subtilis* ortholog it features a kinase domain[42] which has been shown to function as an arginine kinase in *B. subtilis*. This kinase phosphorylates arginine residues in the DNA binding domain of CtsR and, thereby, inhibits its repressor function and releases it from its target DNA[43]. Beside the described high sequence similarities between the polypeptides encoded by the *B. subtilis* and *C. acetobutylicum clpC* operon and the same structure of the operon, a promoter with high homology to the consensus sequence of σ^A-dependent promoters of Gram-positive bacteria and a CtsR binding site upstream of CA_C3192 (encoding CtsR) matching exactly the *B. subtilis* binding site are present[6,26] (Fig. 4.2B). This clearly classifies the members of the *C. acetobutylicum clpC* operon as class III *hsp* genes. The same described *C. acetobutylicum* features apply for *C. beijerinckii*. Here, the potential *clpC* operon is located at the beginning of the leading strand of the 6-Mbp genome and comprises the locus tags CBEI_0120, CBEI_0121, CBEI _0122, and CBEI_0123 with highest identity to the known *B. subtilis* CtsR, McsA, McsB, and ClpC polypeptides[19,44] (Fig. 4.1C), a σ^A-dependent promoter and a CtsR binding site (Fig. 4.2B). Recently, CBEI_0124 has been predicted to be part of the *clpC* operon[22].

A complete *C. acetobutylicum* and *C. beijerinckii* genome search for other potential CtsR binding sites using the search algorithm provided by the virtual footprint analyzer[33] resulted in no further class III *hsp* gene identifications, assuming the originally described localization of the *cis* regulatory CtsR binding site in the promoter region of the DNA strand to be transcribed[6]. A formerly described CtsR binding site in front of *C. acetobutylicum* hsp18[17, 45] (CA_C3714) is located 169 bp upstream of the start ATG in the reverse direction on the opposite, non-transcribed DNA strand. The same situation was found in front of the *C. beijerinckii* orthologous gene *hsp20* (CBEI_4123, 44% sequence similarity to the deduced amino acid sequence of *C. acetobutylicum HSP18*). Here, the reverse complementary CtsR binding site is positioned 58 bp upstream of the start ATG. Therefore, a regulatory influence of CtsR on these genes has to be investigated, although elevated levels of *C. acetobutylicum* HSP18 or its RNA in response to heat[45,46], metabolite stress[12,17], and the onset of solventogenesis[11,45] have been shown.

In *B. subtilis*, all members of the *clpC* operon are involved in the autoregulation of their own expression[42,43,47,48] as well as the expression of *clpP* and *clpE*[7]. ClpP and ClpE are, like ClpC, representatives of the ATP-dependent Clp-family of proteases and most likely involved in the regulation of the CtsR titer[42,49]. Thereby, heat- and general stress-inducible ClpP[50] functions as a proteolytic subunit, ClpE and ClpC as ATP binding units[4,10]. The non-identification of CtsR binding sites in the promoter region of homologous Clp-family proteases in *C. acetobutylicum* and *C. beijerinckii* might, compared to *B. subtilis*, result in a different activity control of the autoregulator CtsR in the two clostridia.

4.5 Class IV–VI heat shock proteins

Class IV *hsp* genes are generally genes whose expression is regulated by an unknown mechanism or a different mechanism to the one described, but clearly respond to heat shock and other stress factors[8,9]. For *B. subtilis* a more detailed classification of these *hsp* genes into a class IV *htpG*-operon, a class V CssSR regulon comprising the membrane-bound proteases *htrA* and *htrB*, and a class VI summarizing stress responsive genes with unknown regulatory mechanisms, has been suggested[10].

A stress-inducible *htpG* (CA_C3315) *hsp* gene has recently been described in *C. acetobutylicum*. However, due to the possession of a CIRCE element in its promoter region it was assigned to the class I *hsp* genes[12,17]. The *C. beijerinckii htpG* (CBEI_4160, see above) does not possess such a CIRCE element and, therefore, does not belong to the class I *hsp* genes. But, it does not possess the suggested regulatory *B. subtilis htpG*-operon site GAAAGG either and might, therefore, be effected by another regulation.

A metabolite stress-induced *htrA* gene (CA_C2433) encoding for a HtrA-like serine protease has been identified in *C. acetobutylicum*[12,17]. The *C. beijerinckii* HtrA candidate with the highest sequence identity (42%) to the deduced amino acid sequence of CA_C2433 is a 2-alkenal reductase (Table 4.1, CBEI_0379). Like its *C. acetobutylicum* counterpart it contains a PDZ domain[51,52], a structural part of trypsin-like serine proteases which are often involved in heat shock response, chaperone function, and apoptosis[53]. HtrA proteases are ubiquitous, heat shock inducible and involved in the degradation of misfolded proteins, especially in the periplasm[54]. The *B. subtilis htrA* and *htrB* are strictly controlled by a CssSR two-component system which is involved in the regulation of secretion stress[8,55]. Both *C. acetobutylicum* and *C. beijerinckii* possess potential candidates for the sensor histidine kinase CssS (CA_C3662, CBEI_1732) and the response regulator CssR (CA_C3663, CBEI_1731)[19,20,56]. A regulation of *C. acetobutylicum* and

C. beijerinckii htrA by the potential CssSR systems as well as an induction by heat needs to be investigated.

Little is known about other potential HSPs in *C. acetobutylicum* and *C. beijerinckii*. Data regarding the heat shock response of *C. acetobutylicum* often lack the clear identification of detected HSPs beside DnaK/J, GrpE, GroESL, HSP18, HSP90, and their respective genes[46,57,58]. Genes which have definitely been identified and shown to be induced by metabolite stress are, e.g. *clpP*, *clpX*, *clpA*, and *lonA*[12,17]. To recap, in *B. subtilis*, *clpP* and *clpC* (greatest similarity to *C. acetobutylicum clpA*[19]) are members of the *CtsR* class III *hsp* genes. In *C. acetobutylicum* and *C. beijerinckii*, besides *clpC*, none of the *clp* genes possess a CtsR binding site in its promoter region (see above). Of the other proteases, *clpX* (CA_C2639) encodes a member of the highly conserved, ubiquitous ATP-dependent proteases[4,20]. It has been shown that ClpX alone possesses all the characteristics of a molecular chaperone and is induced by heat shock[59,60]. In *B. subtilis* it is involved in several cellular processes[61]. Together with ClpP, a proteolytic subunit of ATP-dependent Clp-proteases, it acts in the degradation of misfolded proteins and CtsR, the transcriptional repressor of class III *hsp* genes[38,42]. The *C. beijerinckii* homolog is encoded by CBEI_1328[19] (Table 4.1). Like Clp-family proteases, Lon- and FtsH-family proteases are HSPs with an unknown regulatory mechanism of expression. They belong to the ATP-dependent AAA+ protein superfamily as well and are, like Clp-family proteases, involved in a wide variety of cellular processes including stress response and degradation of misfolded proteins[40,41]. Depending on their active site sequence, heat shock inducible[62,63] Lon-family proteases are separated into two subfamilies, LonA and LonB[64]. The genomes of *C. acetobutylicum* and *C. beijerinckii* contain the genetic information for two LonA proteases and two LonB proteases, respectively (Table 4.1). For *C. acetobutylicum*, an induction of the annotated LonA protease encoded by CA_C0456 under metabolite stress as well as the initially higher expression of the LonA protease encoded by CA_C2637 under butanol stress have been shown[11,17]. One interesting feature in the case of *C. acetobutylicum* and *C. beijerinckii* is the localization of the genes of one LonA protease and one LonB protease next to each other (CA_C2637, CA_C2638, and CBEI_1330,

CBEI_1329) and in close proximity to *clpX*, *clpP* and *engB* which potentially codes for a ribosome biogenesis GTP-binding protein (YsxC). Whilst *clpP*, *clpX*, *lonB*, *lonA*, and *engB* are localized on the lagging strand, in the *C. acetobutylicum* genome (CA_C2640, CA_C2639, CA_C2638, CA_C2637, CA_C2636), they are localized on the leading strand in the *C. beijerinckii* genome (CBEI_1327, CBEI_1328, CBEI_1329, CBEI_1330, CBEI_1331). The homology values of the deduced amino acid sequences of the LonA proteases and LonB proteases are shown in Table 4.1. The potential EngB (CA_C2636) of *C. acetobutylicum* exhibits a 68% similarity to the potential EngB of *C. beijerinckii* (CBEI_1331). In *B. subtilis*, *clpX* (BSU28220), *lonB* (BSU28210), and *lonA* (BSU28200) are located next to each other as well, and in *E. coli*, *clpP* and *clpX* even form an operon[59]. Of the FtsH-family proteases, the E. coli FtsH has been shown to be heat and stress induced and involved in the degradation control of σ^{32}, the sigma factor responsible for the transcription of heat shock proteins[65,66]. In *B. subtilis* the same protease is not involved in heat shock regulation, but is important for growth under elevated temperatures and salt concentrations[67]. Based on gene expression patterns a metabolite stress induction of *C. acetobutylicum ftsH*[20] (CA_C0602, CA_C3202) could not be shown[12,17,20].

Besides the described HSPs, other polypeptides with a potential chaperone or protease activity are present in *C. acetobutylicum* and *C. beijerinckii*. For example, in addition to the class I *dnaK* (CA_C1282) gene, two further genes encoding DnaK-like proteins (CA_C0472, Ca_C0473) are annotated in the genome of *C. acetobutylicum* (Table 4.1). These have not been described as HSPs yet but are, due to their sequence homology to the main DnaK[19], (Table 4.1) potential HSP candidates.

4.6 Heat shock proteins (HSPs) and solvent tolerance

Both *C. acetobutylicum* and *C. beijerinckii* produce butanol and are of special interest in the commercial-scale production of this solvent[68-72]. However, one major drawback is the toxicity of the solvent itself. For *C. acetobutylicum* solvent concentrations exceeding more than 1.5–2%, (v/v) have been shown to inevitably lead to growth inhibition and cell death[73–75]. But, over the years it became evident that HSPs play an important role in solvent stress and tolerance. The induction of HSPs, particularly of class I, upon the addition of butanol to *C. acetobutylicum* wild-type and derived strains was demonstrated in several studies[11,12,17]. Moreover, overexpression of GroESL led to a reduced growth inhibition, a prolonged cell metabolism and higher solvent titers[18]. In *C. beijerinckii*, little is known about the expression of HSPs during growth and increasing solvent concentrations. Solvent tolerance was reported to be associated with a downregulation of a glycerol dehydrogenase (*gldA*)[76], and elevated levels of glycosylated cell membrane proteins[77]. However, a connection between HSPs and solvent stress and tolerance has been shown for several organisms, e.g. *Lactococcus lactis*, *Lactobacillus plantarum*, *Streptomyces cerevisiae*, *Pseudomonas putida*, and *E. coli*[78-83], and, therefore, this most likely applies to *C. beijerinckii* as well.

4.7 Conclusions

The genome-wide comparison of the genes encoding the general stress proteins of *C. acetobutylicum* and *C. beijerinckii* has shown high similarities regarding the presence of well-known heat shock protein (*hsp*) genes, their classification into the described *B. subtilis hsp* gene categories, as well as their organization within operons and their regulatory structures. To summarize, *C. acetobutylicum* and *C. beijerinckii* possess class I, III, and IV *hsp* genes. Known class I representatives are *dnaK*, *dnaJ*, *grpE*, *groES*, and *groEL*. *C. acetobutylicum hsp90* is a member of class I *hsp* genes, but the only potential *C. beijerinckii* homolog (CBEI_4130) does not possess class I *hsp* gene characteristics. Both in *C. acetobutylicum* as well as in *C. beijerinckii*, class III *hsp* genes include the members of the *clpC* operon. A CtsR binding site upstream of *hsp18* (*C. acetobutylicum*) or *hsp20* (*C. beijerinckii*) is located in a reverse complementary direction regarding the start ATG. No CtsR binding site could be found upstream of *clpP* and *clpE* representatives as has been shown for *B. subtilis*[6]. Both organisms possess genes for potential class IV–VI candidates (e.g. *htrA*, *lonA, lonB, clpX*, Table 4.1), however, they do not appear to contain class II type *hsp* genes.

With the availability of the genome sequences of both organisms, future analysis of specific stress condition by, e.g. transcriptomics and/or proteomics, will surely contribute to a better understanding of the stress response in the solventogenic clostridia.

Acknowledgement

This work was supported by the Netherlands Organisation for Scientific Research (NWO) through the SysMO project COSMIC1 (826.06.003).

References

1. Morimoto R. I. *et al.*, *The Biology of Heat Shock Proteins and Molecular Chaperones* (Cold Spring Harbor Laboratory Press, Cold Spring Harbor, NY, 994).
2. Tatusov R. L. *et al.*, *BMC Bioinformatics* 4 (2003).
3. Tatusov R. L. *et al.*, *Nucl. Acids Res.* 28 (2000).
4. Gottesman S., *Annu. Rev. Genet.* 30 (1996).
5. Hecker M. *et al., Mol. Microbiol.* 19 (1996).
6. Derré I. *et al., Mol. Microbiol.* 31 (1999).
7. Derré I. *et al., Mol. Microbiol.* 32 (1999).
8. Hyyryläinen H. -L. *et al., Mol. Microbiol.* 41 (2001).
9. Schumann W., *et al.,* in *Bacillus subtilis and Its Closest Relatives: From Genes to Cells,* Sonenshein A., Hoch J. and Losick R., Eds (ASM Press, Washington DC, 2002), pp. 359–368.
10. Schumann W., *Cell Stress Chaperon.* 8 (2003).
11. Alsaker K. V. and Papoutsakis E. T., *J. Bacteriol.* 187 (2005).
12. Alsaker K. V. *et al., Biotechnol. Bioeng.* 105 (2010).
13. Collins M. D. *et al., Int. J. Syst. Bacteriol.* 44 (1994).
14. Slepecky R. A. and Hemphill H. E., in *The Prokaryotes: A Handbook on the Biology of Bacteria*, Vol. 4, 3rd Edition, Dworkin M., *et al.*, Eds (Springer, Singapore, 2006), pp. 530–562.
15. Narberhaus F. and Bahl H., *J. Bacteriol.* 174 (1992).
16. Narberhaus F. *et al., J. Bacteriol.* 174 (1992).
17. Tomas C. A., *et al., J. Bacteriol.* 186 (2004).
18. Tomas C. A. *et al., Appl. Environ. Microbiol.* 69 (2003).
19. Altschul S. F. *et al., Nucl. Acids Res.* 25 (1997).
20. Nölling J. *et al., J. Bacteriol.* 183 (2001).

21. Copeland A. *et al.*, *US DOE Joint Genome Institute* NCBI Accession Number NC_009617 (2007).
22. Wang Y. *et al., BMC Genomics* 12 (2011).
23. D'aubenton Carafa Y. *et al., J. Mol. Biol.* 216 (1990).
24. Lesnik E. A. *et al., Nucl. Acids Res.* 29 (2001).
25. Platt T., *Annu. Rev. Biochem.* 55 (1986).
26. Sonenshein A. L. *et al.*, in *Handbook on Clostridia,* Dürre P., Ed. (Taylor and Francis, London, 2005), pp. 607–630.
27. Zuker M., *Nucl. Acids Res.* 31 (2003).
28. Homuth G. *et al., J. Bacteriol.* 179 (1997).
29. Fischer R. -J. *et al., J. Bacteriol.* 175 (1993).
30. Fischer R. -J. *et al., J. Bacteriol.* 188 (2006).
31. Nair R. V. *et al., J. Bacteriol.* 181 (1999).
32. Versteeg S. *et al., J. Bacertiol.* 185 (2003).
33. Münch R. *et al., Bioinformatics* 21 (2005).
34. Helmann J. D. *et al., J. Bacteriol.* 183 (2001).
35. Krüger E. *et al., Mol. Microbiol.* 20 (1996).
36. Krüger E. *et al., Microbiology* 143 (1997).
37. Krüger E., *et al., J. Bacteriol.* 176 (1994).
38. Krüger E. *et al., J. Bacteriol.* 182 (2000).
39. Zuber P., *J. Microbiol.* 38 (2000).
40. Ogura T. and Wilkinson A. J., *Genes Cells* 6 (2001).
41. Sauer R. T. *et al., Cell* 119 (2004).
42. Krüger E. *et al., EMBO J.* 20 (2001).
43. Fuhrmann J. *et al., Science* 324 (2009).
44. Krüger E. and Hecker M., *J. Bacteriol.* 180 (1998).
45. Sauer U. and Dürre P., *J. Bacteriol.* 175 (1993).
46. Pich A. *et al., Appl. Microbiol. Biotechnol.* 33 (1990).
47. Kirstein J. and Turgay K., *J. Mol. Microbiol. Biotechnol.* 9 (2005).
48. Kirstein J. *et al., EMBO J.* 24 (2005).
49. Miethke M. *et al., J. Bacteriol.* 188 (2006).
50. Gerth U. *et al., Mol. Microbiol.* 28 (1998).
51. Finn R. D. *et al., Nucl. Acids Res.* 38 (2010).
52. Markowitz V. M. *et al., Nucl. Acids Res.* 34 (2006).
53. Clausen T. *et al., Mol. Cell* 10 (2002).
54. Pallen M. J. and Wren B. W., *Mol. Microbiol,* 26 (1997).
55. Horvátha I. *et al., Biochim. Biophys. Acta.* 1778 (2008).
56. Kanehisa M. *et al., Nucl. Acids Res.* 32 (2004).
57. Bahl H. *et al., FEMS Microbiol. Rev.* 17 (1995).
58. Rüngeling E., *et al., FEMS Microbiol. Lett.* 170 (1999).
59. Gottesman S. *et al., J. Biol. Chem.* 268 (1993).
60. Wawrzynow A. *et al., EMBO J.* 14 (1995).

61. Nanamiya H. *et al.*, *J. Biochem.* 133 (2003).
62. Phillips T. A. *et al.*, *J. Bacteriol.* 159 (1984).
63. Riethdorf S. *et al.*, *J. Bacteriol.* 176 (1994).
64. Rotanova T. V. *et al.*, *Eur. J. Biochem.* 271 (2004).
65. Herman C. *et al.*, *P. Natl. Acad. Sci. USA* 92 (1995).
66. Tomoyasu T. *et al.*, *EMBO J.* 14 (1995).
67. Deuerling E. *et al.*, *Mol. Microbiol.* 23 (1997).
68. Dürre P., *Biotechnol. J.* 2 (2007).
69. Dürre P., *Ann. N. Y. Acad. Sci.* 1125 (2008).
70. Festel G. W., *Chem. Eng. Technol.* 31 (2008).
71. Jang Y. S. *et al.*, *Biotechnol. J.* 7 (2012).
72. Tracy B. P. *et al.*, *Curr. Opin. Biotechnol.* 23 (2011).
73. Lin Y. -L. and Blaschek H. P., *Appl. Environ. Microbiol.* 45 (1983).
74. Ounine K. *et al.*, *Appl. Environ. Microbiol.* 49 (1985).
75. Vollherbst-Schneck K. *et al.*, *Appl. Environ. Microbiol.* 47 (1984).
76. Liyanage H. *et al.*, *J. Mol. Microbiol. Biotechnol.* 2 (2000).
77. Wang F. *et al.*, *Microbiol.* 151 (2005).
78. Desmond C. *et al.*, *Appl. Environ. Microbiol.* 70 (2004).
79. Fiocco D. *et al.*, *Appl. Microbiol. Biotechnol.* 77 (2007).
80. Kang H. J. *et al.*, *Biochem. Biophys. Res. Commun.* 358 (2007).
81. Segura A. *et al.*, *J. Bacteriol.* 187 (2005).
82. Vianna C. R. *et al.*, *Antonie Van Leeuwenhoek* 93 (2008).
83. Volkers R. J. *et al.*, *Environ. Microbiol.* 8 (2006).

CHAPTER 5

MATHEMATICAL MODELING OF THE PH-INDUCED METABOLIC SHIFT IN *CLOSTRIDIUM ACETOBUTYLICUM*

T. MILLAT, S. HAUS, O. WOLKENHAUER

Department of Systems Biology & Bioinformatics
University of Rostock 18051 Rostock, Germany

The Gram-positive bacterium *Clostridium acetobutylicum* responds to variations of the external pH value by changing its metabolic profile. Since it is unable to maintain a constant internal pH value it shifts its metabolism between two distinct phases differing in metabolome, proteome, and transcriptome. During acidogenesis (high pH value) *C. acetobutylicum* primarily produces the acids acetate and butyrate. In contrast, the bacterium enters solventogenesis for low pH levels during which the main products are the industrially relevant solvents acetone and butanol.

This solventogenic shift is a dynamic process governed by changes in the composition of transcriptome, proteome, and metabolome. Here, we present a model that considers the transition from acidogenesis to solventogenesis in a continuous culture. The external pH value is regulated by the experimental set-up and can be shifted in both directions. The model parameters are estimated from available dynamic shift experiments. Finally, the model is compared to these experiments.

The modeling approach reveals that the metabolic shift in *C. acetobutylicum* involves several levels of cellular organization. Thus, the present model includes biochemical, environmental, transcriptional, proteomic, and metabolomic information. In particular we show that the consideration of the pH-dependency of cellular processes is crucial for the modeling of the pH-induced metabolic shift in *C. acetobutylicum*.

5.1 Industrial application of *C. acetobutylicum*

Since Louis Pasteur first observed that bacteria can produce butanol in 1861, there has been an interest in the underlying fermentation process from both scientific and biotechnological perspectives. In 1912, Chaim

Weizmann discovered that the Gram-positive bacterium *C. acetobutylicum* is capable of fermenting starchy materials to acetone, butanol, and ethanol. Shortly after this discovery, in 1918, the first commercial clostridial fermentation plant was built in Terra Haute, Indiana. Twenty years later, after the expiration of Weizmann's patent, plants were built worldwide and clostridial fermentation became a major industrial fermentation process.

This development was stopped in the early 1960s when the production of these chemicals from petrochemicals became more economically favorable. The interest was renewed in the late 20th century as a consequence of both rising costs for crude oil and natural gas and the energy crisis. However, even with the tremendous progress in the development of genetic systems used to improve the fermentative capabilities of *C. acetobutylicum*, the industrial clostridial fermentation did not reach its former level.

With the beginning of the new millennium the situation changed for different reasons[1]. First, the depletion of fossil-fuel reserves challenged the future energy systems worldwide. In 2009, the world's energy consumption was approximately 339 EJ (1 EJ=10^{18} J), and some 80.9% of this came from the burning of fossil fuels[2]. Second, the prices for fossil fuels increased, and thus clostridial fermentation became economically favorable once more. Third, climate change as a result of the extensive use of fossil fuels demanded alternative sources for energy and fuel with better ecological compatibility. These economic and ecologic restrictions gave rise to a renewed interest in alternative sources of energy and chemicals. However, future energy production has to fulfill several demands. First, the substitutes must match large-scale production without adding to the accumulation of CO_2 in the atmosphere[1]. The clostridial fermentation of butanol may satisfy this condition. Second, the alternative energy sources should not compete with food production. Third, they should provide high-density, transportable, and manageable energy. Here, butanol is superior to ethanol due to its higher energy density and its more favorable chemical and physical properties[3].

Despite the past industrial use of clostridial fermentation, future applications require enhanced production of fermentation products, e.g.

butanol, per glucose equivalent, and an increased production rate. These prerequisites necessitate an improved understanding of the cellular processes and their regulation, the interactions between the microbial colony and its environment, and key parameters for external control. For this purpose, data-driven mathematical models have to be established allowing for the systematic optimization of clostridial fermentation.

In this chapter, we present a model of the pH-induced metabolic shift from acidogenesis to solventogenesis in *C. acetobutylicum*. In particular, we focus on model generation using existing biological information as well as data originating from recent model-driven experiments. To this end, we briefly introduce the metabolic network in the next section. The following main section is divided into four parts focusing on different aspects of the model. First, we discuss the standardized experimental set-up used and its consequences for the model. Second, based on emerging experimental evidences, we expatiate on the relationship between changes on transcriptome and proteome levels. Third, we discuss the influence of pH-dependent enzyme kinetic reaction on the metabolic shift in *C. acetobutylicum*. Finally, we combine the previous information into a dynamic model of the pH-induced shift and compare the simulation of the model to experimental data. Afterwards, we summarize our conclusions and give an outlook for the future modeling of this pH-induced metabolic shift.

5.2 AB fermentation in *C. acetobuylicum*

The acetone-butanol fermentation (AB fermentation), which is also referred to as acetone-butanol-ethanol (ABE) fermentation, is a notable feature of the metabolism of *C. acetobutylicum*. This microbial fermentation is distinguished by two different metabolic phases — acidogenesis and solventogenesis. In response to varying pH levels, the bacteria shift between both phases. It is believed that the ability of *C. acetobutylicum* to do this is one aspect of its pH-induced stress response[3,4]. The metabolic shift may delay entry into sporulation and, thus, give the clostridial population an advantage over other bacteria in the same ecological niche.

In a continuous culture under phosphate limitation, both phases are characterized by different external and internal pH levels. During acidogenesis (pH > 5.2) the bacteria predominately produce the acids acetate and butyrate, whereas the solvents acetone and butanol are the main products during solventogenesis (pH < 5.1)[5-8]. Furthermore, the cells generate ethanol in similar amounts during both metabolic phases. Figure 5.1 summarizes these experimental findings. It displays the steady-state concentrations of the fermentation products as a function of the external pH level. Furthermore, we used hyperbolic tangents to fit the experimental data[9]. The inflection points of these functions were used to estimate the phase limits of acidogenesis and solventogenesis.

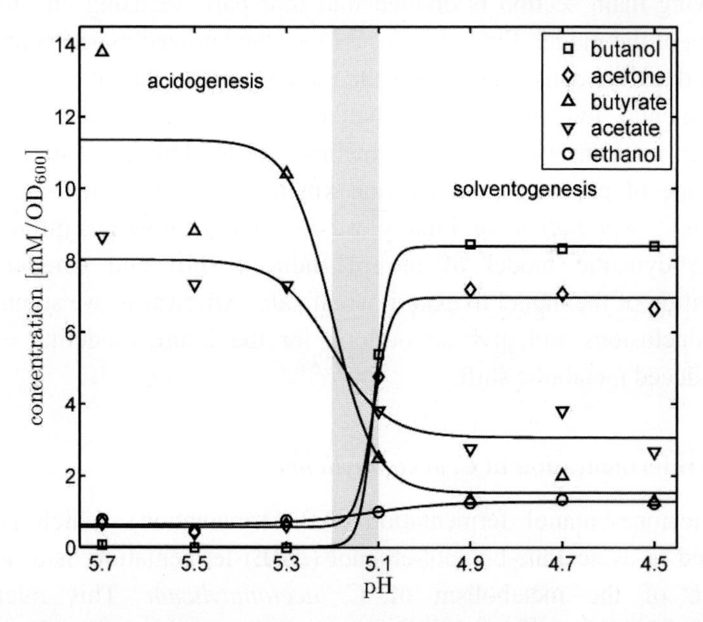

Fig. 5.1: The steady-state concentration of the fermentation products in *C. acetobutylicum* as a function of the external pH level. The symbols denote recent experimental data[10] using continuous cultures under phosphate limitation[10,11]. We normalized the data using the optical density (OD_{600}) to account for variations in the population size. Hyperbolic tangents are used to fit the data (solid lines). Their inflection points define the limits of acidogenic, solventogenic, and transition phases (gray). The acids butyrate and acetate are predominantly fermented during acidogenesis (high pH). During solventogenesis (low pH value) the solvents butanol and acetone are the main metabolic products. In the pH range from 5.2 to 5.1, the bacteria perform a dramatic metabolic phase shift.

Interestingly, we found that a transition phase (5.1 < pH < 5.2) emerges which cannot be assigned to acid or solvent production. During this transition phase, *C. acetobutylicum* dramatically alters its metabolic profile.

The fermentative metabolism of *C. acetobutylicum* has been investigated over several decades. In fact, this bacterium is the best-studied solventogenic *Clostridium*. Thus, the metabolic pathway of AB fermentation is well known[4], see Fig. 5.2.

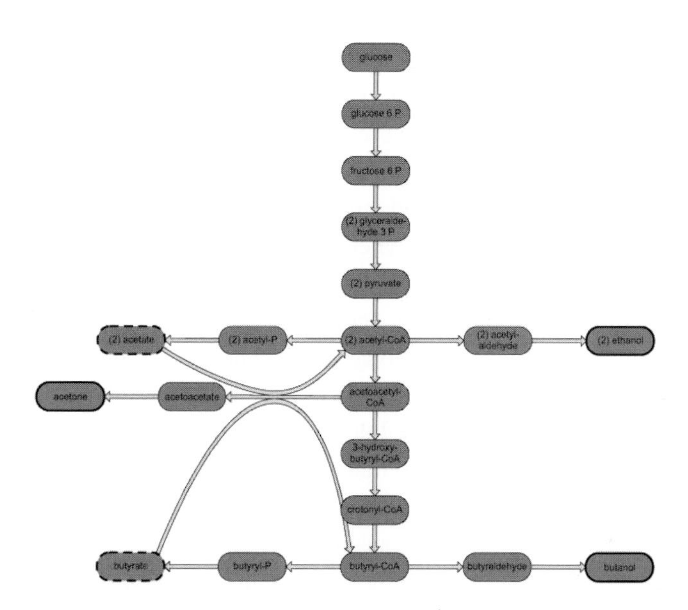

Fig. 5.2: The metabolic pathway of AB fermentation[4]. Dashed edges indicate the acids (acetate and butyrate), which are the main products during acidogenesis. The solvents are denoted with solid edges. Acetone and butanol are mainly produced during solventogenesis. Ethanol is fermented in approximately the same amounts in both metabolic phases. The pH-dependent steady-state concentrations are shown in Fig. 5.1.

As a first step, extracellular glucose is transported into the cell by a phosphoenolpyruvate-dependent phosphotransferase system[12]. Subsequently, pyruvate is formed in several glycolytic steps[4]. Pyruvate is further converted to acetyl-CoA, which is an important branch-point in the AB fermentation. It is the starting point for the formation of acetate and ethanol[13]. From the second branch-point in AB fermentation, acetoacetyl-CoA, acetone is generated by two sequential enzyme catalytic reactions with the intermediate acetoacetate[14].

Interestingly, the enzymes involved in this conversion, acetoacetyl-CoA transferase and acetoacetate decarboxylase, are arranged in two different operons[14]. This suggests that their regulation is mediated by different mechanisms. Finally, glucose is reduced to butyryl-CoA. This last branch-point of the AB network is the starting point for the formation of butyrate and butanol[13].

During acidogenesis, acetyl-CoA and butyryl-CoA are converted to acetyl- and butyryl-phosphate. Then, these acylphosphates are transformed to acetate and butyrate (see Jones and Woods (1986)[4] for a review). Afterwards, the acids are excreted from the cell. During the transition from acid to solvent production, these acids are re-absorbed by the cells[15,16]. The reinternalized butyrate and acetate are converted to butyryl-CoA and acetyl-CoA, where both reactions are mediated by a CoA-transferase[15,16] that consists of the subunits CtfA/B[17].

During the solventogenesis, acetylaldehyde and butyraldehyde are formed from acetyl-CoA and butyryl-CoA. Then, these intermediates are converted to ethanol and butanol. To accomplish these reductions, a bifunctional acetaldehyde/alcohol dehydrogenase (AdhE) is required[18].

Recent experiments indicate that the transition from acidogenesis to solventogenesis involves transcriptomic, proteomic, metabolomic, and environmental changes[8,19]. In consequence, a model of the pH-induced metabolic shift has to consider these different biological levels. Furthermore, the experimental set-up and the impact of the changing pH level on enzymatic reactions have to be taken into account.

Based on this experimental information, we establish a mathematical model in the next section. To this end, we first consider several aspects of the pH-induced metabolic shift in isolation. Afterwards, we combine them into a joint model of this phenomenom.

5.3 Modeling the pH-dependent metabolic switch

The pH-induced metabolic shift in *C. acetobutylicum* involves several levels of biological organization. Consequently, the modeling requires a sophisticated knowledge and experimental information of these levels. However, the dynamic measurement of transcriptomic, proteomic, enzymatic, metabolic, and environmental changes is usually beyond the capacities of a single lab[20]. Thus, different labs with different specializations have to collaborate closely in order to provide data for modeling. Nevertheless, the challenge is to establish an experimental set-up and standard operating procedures over all collaborating labs that fit the needs of the planned measurements and the modeling.

To illustrate the demands that emerge from modeling, we first discuss the ingredients for a model of the pH-induced metabolic shift. Afterwards, we combine them into a dynamic model.

5.3.1 The AB fermentation in a continuously fed well-stirred isothermal tank reactor

To investigate the pH-induced metabolic shift, a continuous culture under phosphate limitation was chosen using the previously described standard operating procedures[11,21]. This experimental set-up is superior to traditionally used batch cultures, because it reduces the environmental and biological factors influencing the bacterial population. Thus, it allows for a reasonable restriction of processes considered in the model. However, clostridial populations may behave substantially differently under continuous and batch culture conditions[19].

In the chosen continuous culture set-up, the external pH level is the only environmental variable changed during the experiments. The continuous culture is fed with a constant inflow of glucose (4% glucose). Simultaneously, final metabolic products are removed from the system with the same dilution rate. Previous experiments observed that the solvent production depends on the dilution rate[6]. Based on these results, the dilution rate D was chosen to maximize the solvent yield. Furthermore, the system is well stirred to guarantee a homogenous distribution of the microbial population in the vessel. This set-up is referred to a continuously fed well-stirred isothermal tank reactor (CSTR) and is an example for open systems[22].

To describe the dynamics of the concentrations in such a reactor, one has to consider the contributions from the reaction kinetics, e.g. the conversion of glucose into butanol, and from the in- and outflow of the substances. To illustrate the mathematical representation, we consider the internal glucose concentration. Its rate of change is represented as

$$\frac{dGlc}{dt} = D \cdot Glc_0 - (k_1 + D)Glc \tag{5.1}$$

where Glc_0 is the glucose concentration in the stream of liquid feeding the reactor and Glc the glucose concentration in the reactor. For the sake of simplicity, we represented the conversion of glucose by a monomolecular reaction. In the above rate equation, positive terms describe processes adding glucose and negative terms describe processes decreasing the concentration either by biochemical conversion or due to the constant in- and outflow. Furthermore, the parameter k_1 is the kinetic coefficient of the metabolic conversion of glucose. This example illustrates that, in continuous cultures, the concentrations of metabolites depend on biochemical (metabolic) processes and the flow of substances.

For final metabolic products, e.g. butanol, the rate equation simplifies to

$$\frac{dB}{dt} = k \cdot X - D \cdot B \tag{5.2}$$

because there is no inflow of final products into the cell. Here, $\mathrm{k} \cdot \mathrm{X}$ exemplarily stands for any conversion process of an intermediary metabolite resulting in butanol.

In this section, we demonstrated that the experimental design impacts the modeling framework and, naturally, the experimental data. In consequence, modeling starts with the description of the set-up used for the experiments. The consideration of a CSTR adds a transport term to the rates of the metabolites and intermediates. This addition may seem to be simple, but in fact it is crucial for the system's behavior, because it changes dynamics and steady states of the system. In general, non-linear complex behavior may emerge in open systems which cannot be observed in closed systems[22].

Furthermore, the experimental design has to be appropriate to provide reliable and reproducible information for the model. The standardized continuous culture under phosphate limitation and the pH as the sole external variable was chosen for that purpose. Based on this experimental set-up, multiple experiments have been conducted providing information for the model of the pH-induced metabolic shift, which we present in the subsequent sections.

5.3.2 pH-induced changes of transcriptome and proteome

During the pH-induced metabolic shift, *C. acetobutylicum* alters its transcriptomic[19] and proteomic profile[8]. Consequently, we have to incorporate the dynamic relationship between the amount of gene transcript and encoded enzyme in an appropriate manner. To this end, we express the protein concentration as

$$\frac{dP}{dt} = k_E \cdot mRNA - k_D \cdot P \tag{5.3}$$

For the sake of simplicity, we assumed that the protein synthesis is proportional to the amount of the messenger RNA encoding the protein. The kinetic coefficient k_E describes the averaged probability that a gene is expressed per unit time. Thus, it is determined by the speed of transcription and translation. Furthermore, it contains information about the probability of a successful completion. It also depends on the stability of the messenger RNA which, in turn, may be dependent on the pH value[23]. The intracellular protein degradation is determined by the second term in Eq. (5.3). It is proportional to the degradation coefficient k_D, which is a measure of the protein stability. However, we note that the synthesis of proteins may be also regulated at translational and posttranslational steps[24,25].

In agreement with experimental data[26-28], this simple model illustrates that the dynamic proteome is disproportional to the mRNA level. Instead, its level has to be interpreted as the integral over the time course of the mRNA level. In the further course of this section, we discuss the impact of this important conclusion on the model of the pH-induced shift.

Prior to that, we briefly introduce the steady-state properties of rate Eq. (5.3). At steady state, the synthesis and the degradation of the protein

P are balanced. Consequently, its rate of change is zero and can be transformed into the algebraic equation

$$0 = k_E \cdot mRNA - k_D \cdot P \tag{5.4}$$

so that the steady-state concentration P^{SS} follows as

$$P^{SS} = \frac{k_E}{k_D}\ mRNA \tag{5.5}$$

which is a linear function with respect to the level of the messenger RNA. Its slope is determined by the ratio of kinetic coefficients k_E and k_D. Hence, the protein level cannot be concluded from the mRNA level if this ratio is unknown.

Assuming a stepwise alteration of the amount of transcript, the differential Eq. (5.3) can be solved analytically, Fig. 5.3,

$$P(t) = \frac{k_E}{k_D}\ mRNA \cdot \left(1 - e^{-k_D(t-t_0)}\right) + e^{-k_D(t-t_0)} P_0 \tag{5.6}$$

where the first term describes the approach to the new steady state and the second term the "degradation" of the previous state. Interestingly, the relaxation to the steady state only depends on the degradation coefficient k_D and time t. From Eq. (5.6) we can deduce the relaxation time

$$\tau = \frac{1}{k_D} \tag{5.7}$$

which measures how long it takes to approach the new steady state. The relaxation time is inversely proportional to the degradation rate coefficient and, thus, increases with increasing protein stability. Furthermore, this characteristic time is used to classify the system's ability to respond to changes in the mRNA level. For variations much slower than the relaxation time, the system follows the changes on the mRNA level. On the other hand, faster fluctuations cannot be recognized and will be suppressed on the proteome level.

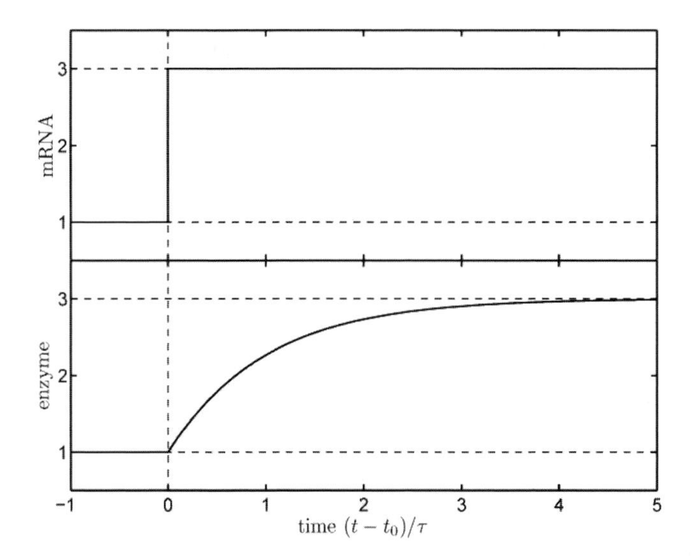

Fig. 5.3: The dynamic transition between two different steady states for a stepwise change of the mRNA level as a function of time. The upper diagram shows the time course of the relative amount of mRNA. The corresponding evolution of the relative protein level is plotted in the lower part.

The simple model (Fig. 5.3) shows that the change in the protein concentration is delayed with respect to the mRNA level. This delay increases further if one considers protein synthesis in more details. Furthermore, it illustrates that the time-dependent behavior of mRNA and protein abundance follows different functions.

The gene expression during the metabolic shift in *C. acetobutylicum* has been measured in recent experiments[19]. These experiments found that the expression of several genes encoding for solvent-forming enzymes is maximally induced during the transition phase and decreases later. Surprisingly, the expression of some genes is not significantly changed during solventogenesis, whereas the protein level is significantly modified[8]. Based on these findings, we expand our model to investigate the response of the protein level to a change of the mRNA level, which exhibits such a behavior.

Because the regulatory mechanisms are still unknown, we use a phenomenological approach and replace the mRNA level in Eq. (5.3) with a time-dependent function. This function starts at the acidogenic mRNA level, increases to a maximum during the transition phase, and tends to a new steady state in solventogenesis, Fig. 5.4. The resulting response of the proteome was calculated numerically and is shown in the lower plot of this Fig. 5.3.

Assuming that the protein degradation is slower than the gene expression, the resulting dynamic protein profile displays a switch-like behavior. In comparison to Fig. 5.3, the new proteomic steady state is approached faster and, the bacterium adapts, thus, quicker to the altered situation. However, the steady-state levels coincide. Similar to a step-like change, it is determined by Eq. (5.5) which states a direct correlation between the steady-state levels of gene expression and protein level.

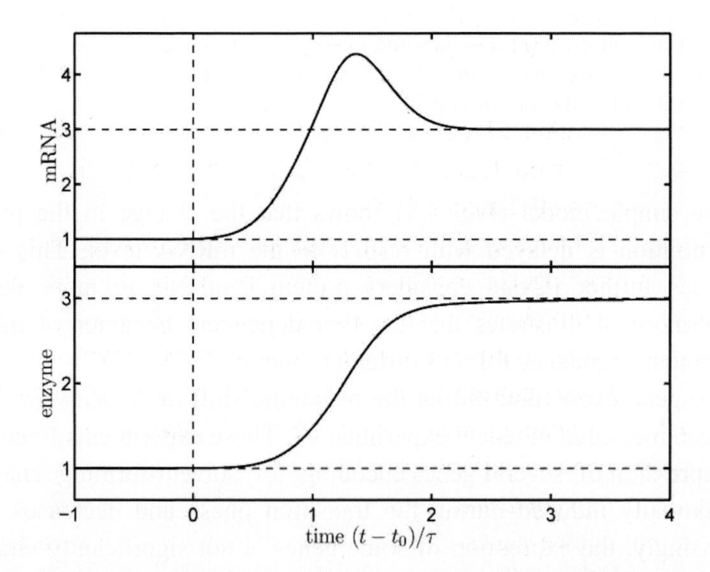

Fig. 5.4: The dynamic transition between acidogenic and solventogenic steady states assuming an mRNA profile recently reported[19]. The upper plot shows the dynamic transcript profile and the lower plot the resulting time-dependent enzyme concentration.

The dynamics of the protein concentration shown in Fig. 5.4 can be approximated by a modified logistic growth function[29,30]:

$$f(t) = \frac{c_1}{2}\left[\tanh\left(\frac{c_2}{2}(t - c_3)\right) + 1\right] + c_4 \tag{5.8}$$

here introduced in an alternative representation. The parameter c_1 defines the range of the function. Its steepness is inversely determined by the parameter c_2. The parameter c_3 specifies the position of the inflection point and, thus, we use it to adjust the curve within the time domain. Furthermore, we introduce the parameter c_4 to account for a basal expression level. In Fig. 5.5, we plot the time course of the logistic growth curve and compare it to characteristic measures. Furthermore, after rearranging Eq. (5.8)[a] into

$$f(t) = \frac{c_1}{2}\left[1 - \tanh\left(\frac{c_2}{2}(t - c_3)\right)\right] + c_4 \tag{5.9}$$

we obtain a function which can be used to describe decreasing protein levels. Except its monotonic decreasing behavior, it shares the properties of the logistic growth function (5.8).

Whereas an increase of solventogenic proteins/enzymes is expected during the solventogenic shift, a universal direct correlation between the transcript and protein levels at steady state was disproven in recent experiments[8,19]. In contrast to that expectation, these experiments have observed that the expression of several mRNAs is not significantly increased in solventogenesis, while the concentration of the corresponding proteins is significantly increased. Similar results were recently reported for *Mycoplasma pneumoniae*[31,32]. Here, the authors even observed that some genes and the encoded proteins behave in an anti-correlated manner; an increased mRNA level had been measured, but the protein abundance was decreasing.

---

[a] A change in the order of the arguments in the hyperbolic tangent from $(t - c_3)$ to $(c_3 - t)$ also results in the same decreasing function. However, we prefer the notation used in Eq. (5.9), because it is more easily distinguished from Eq. (5.8).

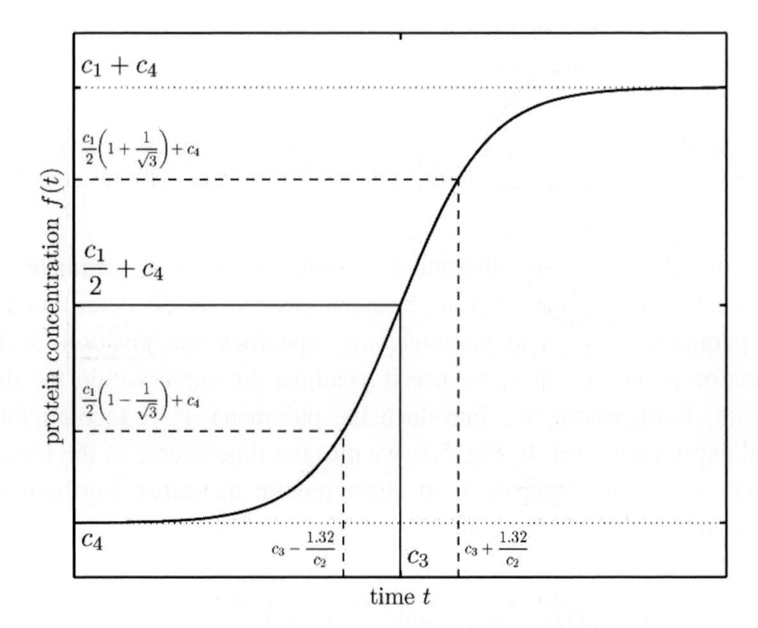

Fig. 5.5: The logistic growth function (5.8) used to model pH-induced changes in enzyme concentrations as a function of time. In addition to the function $f(t)$ (solid line), several characteristic measures are shown which follow from the parameterization.

On the basis of these observations, we add a further regulation to protein synthesis and degradation that reflects experimentally-found changes of transcription factor activity[33], RNA stability[23], protein folding[34], and enzyme activity[35,36]. In Fig. 5.6, we investigate the effect of a pH-dependent degradation coefficient k_D on the dynamic proteome assuming that the protein is more stable during solventogenesis. In contrast to Fig. 5.4, the relative change of the mRNA level differs from that of the protein. Indeed, the protein level quadruplicates due to the pH-dependent protein stability.

As a consequence of these findings, the prediction of protein levels from mRNA levels requires a detailed knowledge of all steps involved in the dynamics of the proteome. Thus, the challenge for future research is to unravel the interrelationships between genomic, posttranscriptional/ translational, and metabolic networks[37]. Probably, the temperature-dependence of kinetic coefficients, commonly expressed by the empirical

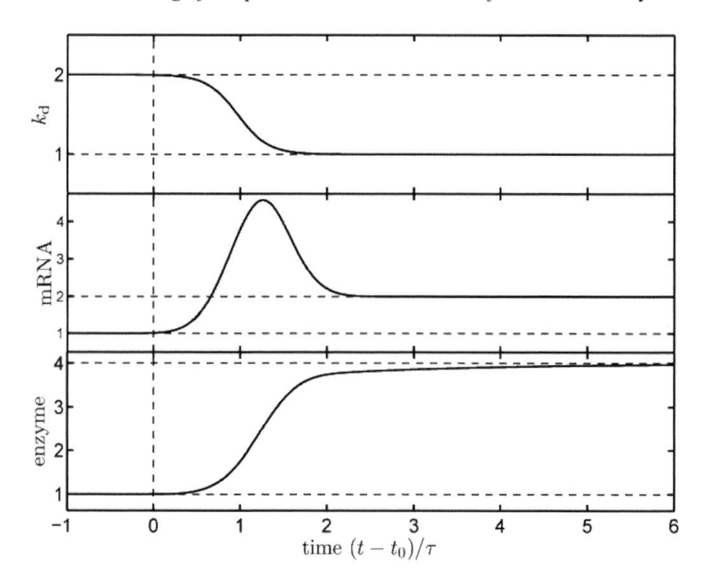

Fig. 5.6: The temporal evolution of the relative protein amount considering a pH-dependent protein synthesis and degradation. In the upper plot, we show the temporal evolution of the pH-dependent degradation coefficient. The mRNA level (middle plot) changes in response to the varying pH. Due to the pH-dependent degradation coefficient, the protein level approaches a steady state that disagrees with the relative change of the mRNA level.

Arrhenius equation[38,39], is the best-known example for the impact of physical parameters on biochemical reactions. However, enzyme kinetic reactions often exhibit a more complex temperature-dependent behavior[35,40,41].

In liquids, solvent and ion (electrolyte) concentrations also impact the reaction rate, which is either referred to a kinetic salt effect[42] or ionic strength effect[39]. (The latter name is preferred in modern literature.) The hydron[b], that is the hydrogen cation $\mathbf{H^+}$, is of particular interest for the modeling of the pH-induced shift. Its concentration is measured by the pH, the sole external variable in the continuous culture experiments.

[b] According to the International Union of Pure and Applied Chemistry (IUPAC), the hydrogen cation H^+ is named in this way[54].

For this reason, we discuss the pH-dependency of biochemical reactions in more detail in the following section.

5.3.3 The impact of the pH value on biochemical reactions

The pH-dependency of enzyme kinetic reactions have been reported since the first experimental studies on enzymatic behavior, e.g. by Michaelis in 1911[43] and 1922[44], but it is neglected in the standard representation of enzyme kinetic reactions

$$\mathrm{E} + \mathrm{S} \underset{k_2}{\overset{k_1}{\rightleftharpoons}} \mathrm{C} \xrightarrow{k_3} \mathrm{P} + \mathrm{E} \tag{5.10}$$

where the enzyme E facilitates the conversion of the substrate S into the product P. Furthermore, an intermediary complex C is formed during this process. Applying several approximations, the Michaelis–Menten equation[35,40,41],

$$\frac{\mathrm{dP}}{\mathrm{dt}} = -\frac{\mathrm{dS}}{\mathrm{dt}} = \frac{\mathrm{V}_{\max} \cdot \mathrm{S}}{\mathrm{K}_{\mathrm{M}} + \mathrm{S}} \tag{5.11}$$

is derived, which is commonly used to model biochemical networks. It is determined by the limiting rate $V_{\max} = k_3 \cdot E^{Tot}$ and the Michaelis–Menten constant $K_{\mathrm{M}} = (k_2 + k_3)/k_1$. We note that the inherent approximations used to derive the Michaelis–Menten equation may affect qualitative and quantitative properties of biochemical networks[45,46].

The assumption of a constant pH is appropriate in organisms that very tightly regulate their intracellular pH. Nevertheless, small changes of 0.1 pH units can have physiological consequences for them and even in those organisms the intracellular pH varies significantly in various cellular compartments[47]. Interestingly, anaerobic bacteria are unable to maintain a constant internal pH, but rather preserve a constant transmembrane pH gradient[7,48]. In particular, batch experiments indicate that *C. acetobutylicum* sustains a pH gradient of approximately $\Delta pH \approx 1$ and follows changes of the external pH without significant delay[7,49]. Thus, we expect that changing pH levels impact the kinetic behavior of the reactions involved in AB fermentation.

To illustrate the effect of varying pH on the efficiency of an enzyme, we use a simple model which assumes that the binding or the dissociation of a hydron shift an enzyme from an active to an inactive state shown in reaction scheme (Eq. (5.12))[36],

$$
\begin{array}{ccccc}
\mathrm{E}^{n+1} & & \mathrm{C}^{n+1} & & \\
\Big\updownarrow{\scriptstyle K_{\mathrm{aE}}} & & \Big\updownarrow{\scriptstyle K_{\mathrm{aC}}} & & \\
\mathrm{S}+\mathrm{E}^{n} & \underset{k_2}{\overset{k_1}{\rightleftharpoons}} & \mathrm{C}^{n} & \xrightarrow{k_3} & \mathrm{E}^{n}+\mathrm{P} \\
\Big\updownarrow{\scriptstyle K_{\mathrm{dE}}} & & \Big\updownarrow{\scriptstyle K_{\mathrm{dC}}} & & \\
\mathrm{E}^{n-1} & & \mathrm{C}^{n-1} & &
\end{array}
\tag{5.12}
$$

where we assume furthermore that the reversible binding of hydrons is much faster than enzymatic reaction. Then, the dissociation constants $K_{\mathbf{a}}$ (association) and $K_{\mathbf{d}}$ (dissociation) are sufficient to describe the (de)protonation of the enzyme and the complex. The additional subscript denotes the enzyme or the intermediary complex, whereas the superscript indicates the number of hydrons bound to the enzyme or the complex.

Then, the rate of this pH-dependent reaction is expressed in an equation formally equivalent to Eq. (5.11). Contrary to the parameters in the Michaelis–Menten (Eq. (5.11)), the apparent limiting rate $\overline{V}_{\mathrm{max}}$ and the apparent Michaelis–Menten constant $\overline{K}_{\mathrm{M}}$,

$$
\overline{\mathrm{V}}_{\mathrm{max}}=\frac{\mathrm{V}_{\mathrm{max}}}{1+\dfrac{\mathrm{H}^{+}}{\mathrm{K}_{\mathrm{aC}}}+\dfrac{\mathrm{K}_{\mathrm{dC}}}{\mathrm{H}^{+}}} \quad \text{and} \quad \overline{\mathrm{K}}_{\mathrm{M}}=\mathrm{K}_{\mathrm{M}}\frac{1+\dfrac{\mathrm{H}^{+}}{\mathrm{K}_{\mathrm{aE}}}+\dfrac{\mathrm{K}_{\mathrm{dE}}}{\mathrm{H}^{+}}}{1+\dfrac{\mathrm{H}^{+}}{\mathrm{K}_{\mathrm{aC}}}+\dfrac{\mathrm{K}_{\mathrm{dC}}}{\mathrm{H}^{+}}}
\tag{5.13}
$$

depend on the pH level. The pH-dependent reaction rate features a typical bell-shaped behavior illustrated in Fig. 5.7 as a function of the pH level and the substrate concentration. The pH-optimum and the half width are determined by the association and dissociation constants[36] (cf. scheme (5.12)). Due to this shape, changes of the intracellular pH may bring about physiological consequences including the pH-induced

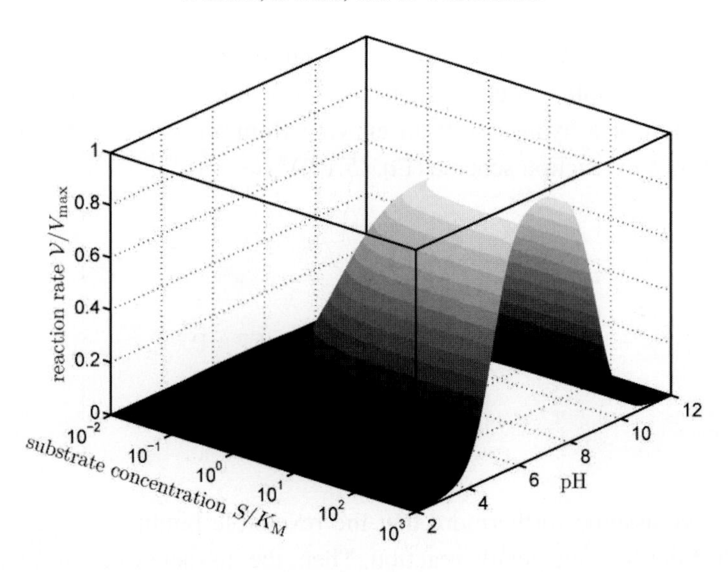

Fig. 5.7: The pH-dependent reaction rate as a function of the pH level and the substrate concentration. For a constant substrate concentration the rate shows a typical bell-shaped behavior if the pH is varied. At constant pH it behaves like the standard Michaelis–Menten equation.

metabolic shift in *C. acetobutylicum*, the torque-speed and rotational direction of the bacterial flagella motor[50], and the pH-dependent effector translocation in *Salmonella enterica*[51]. These observations suggest that microorganisms possess mechanisms to sense the external and/or the internal pH.

Such a sensory element may be a protein which exhibits different states, W and W^*, depicted in reaction scheme (5.14).

$$W \underset{E_2}{\overset{E_1}{\rightleftharpoons}} W^* \tag{5.14}$$

The transition between the states is assumed to be a function of the pH. According to its state, the protein might trigger or suppress gene expression, catalyze a reaction, or activate further cellular responses.

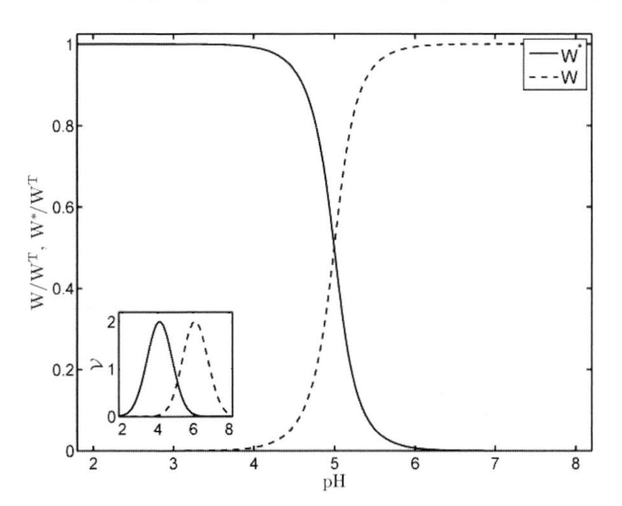

Fig. 5.8: Steady-state concentrations of the distinct states, *W* and *W**, as a function of the pH. The concentrations are normalized using the total protein concentration. The insets display the pH-dependent limiting rates $\boldsymbol{V}_{\mathbf{A}}$ for the transition $W \to W^*$ and $\boldsymbol{V}_{\mathbf{D}}$ for the transition $W^* \to W$. The proportion of the states is pH-independent for coincident pH profiles.

In Fig 5.8, we plot the pH-dependent steady states of the protein forms assuming that the activation operates optimally in acidogenesis, whereas the protein is deactivated during solventogenesis. Because the activation and deactivation of the protein depends on the pH level, the protein manifests a switch-like pH-dependent behavior. Thus, the (de)activation cycle (5.14) may act as a pH-sensory element[9] providing a signal for the cellular adaptation to variations in the extra- and intracellular pH level. Such a stimulus may trigger the induction of solvent-forming enzymes during solventogenesis. Interestingly, the switch-like functions (5.8) and (5.9) could be applied to approximate the pH-dependent activation state of the protein.

In the previous sections, we introduced several aspects of the pH-induced metabolic switch in *C. acetobutylicum* and examined separately the different levels of cellular organization involved in AB fermention. However, experimental evidence has emerged that the levels are closely coupled. For this reason, we combine our findings discussed previously in a dynamic model of the pH-induced metabolic switch and discuss its implications in the next section.

5.3.4 Dynamic modeling of the pH-dependent metabolic shift

The metabolic network of AB fermentation in *C. acetobutylicum* involves more than 20 major conversion steps which include further intermediate reactions. Furthermore, the amount of enzymes facilitating the reactions and their activity depend directly or indirectly on the changing external pH value. Due to the lack of data on the dynamic proteome, we have to simplify the network of AB fermentation. First, we assume that most intermediary steps are in a quasi-stationary state. Due to this simplification, the cascades of reactions can seemingly be reduced into a single step[46]. By this means, we reduced the network of AB fermentation to ten metabolic reactions shown in Fig. 5.9. Second, based on recent experimental data[8,19], we incorporate a pH-dependent induction of the enzymes AdhE, Adc, and CtfA/B. We keep constant the other enzymes involved in AB fermentation during both phases, because new experiments have measured insignificant changes in their steady-state concentration. However, their amount may change significantly during the dynamic shift from acidogenesis to solventogenesis. Third, we apply the logistic functions (5.8) and (5.9) to describe the pH-dependent induction of enzymes. Fourth, we introduce a dilution term to account for the experimental set-up (cf. Section 4.3.1). In addition to the metabolites, we also consider the dilution term for intracellular intermediates and pH-dependent induced enzymes, because living cells are also removed from the chemostat. Fifth, due to a lack of data, we neglect the impact of the pH-level on the enzyme kinetic reactions. In consequence, the limiting rates V_i and the Michaelis–Menten constants K_i are constant parameters that have to be estimated from experimental data.

Based on the reduced model Fig. 5.9, we establish a mathematical model which consists of coupled non-linear differential equations. In Eq. (5.15), we apply nine differential equations to describe the time course of final products and their intermediates (cf. Fig. 5.9).

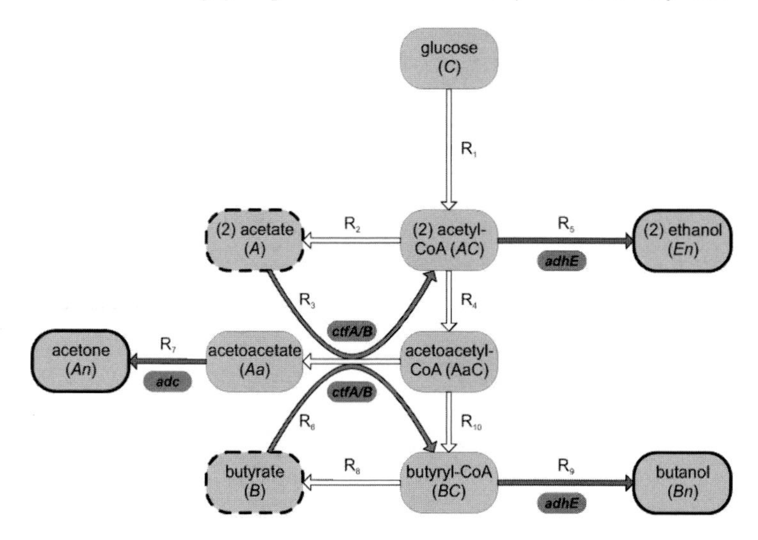

Fig. 5.9: The reduced metabolic pathway of AB fermentation in *C. acetobutylicum*[52]. In contrast to the detailed network represented in Fig. 5.2, most conversions are merged into a single reaction. The reduction results in 10 reactions, R_i. Furthermore, gray arrows denote reactions that are regulated by induction of the facilitating enzymes. Dashed edges represent products of acidogenesis and solid edges products of solventogenesis.

$$\begin{aligned}
\frac{dAC}{dt} &= R_1 - R_2 + R_3 - 2R_4 - R_5 - D \cdot AC \\
\frac{dA}{dt} &= R_2 - R_3 - D \cdot A \\
\frac{dEn}{dt} &= R_5 - D \cdot En \\
\frac{dAaC}{dt} &= R_4 - R_3 - R_6 - R_9 - D \cdot AaC \\
\frac{dAa}{dt} &= R_3 + R_6 - R_7 - D \cdot Aa \\
\frac{dBC}{dt} &= R_9 - R_8 + R_6 - R_{10} - D \cdot BC \\
\frac{dB}{dt} &= R_8 - R_6 - D \cdot B \\
\frac{dAn}{dt} &= R_7 - D \cdot An \\
\frac{dBn}{dt} &= R_9 - D \cdot Bn
\end{aligned} \tag{5.15}$$

Here, the concentration of a component is altered by the reactions R_i which either consume (negative) or produce (positive) this component. Furthermore, a dilution term is added that depends on the dilution rate D and the concentration of the component. The rates of the reactions R_i are:

$$\begin{aligned}
R_1 &= \frac{2V_1 \cdot G}{K_1 + G} & R_6 &= \alpha_6 \cdot B \cdot AaC \cdot Cf \\
R_2 &= \frac{V_2 \cdot AC}{K_2 + AC} & R_7 &= \alpha_7 \cdot Aa \cdot Ad \\
R_3 &= \alpha_3 \cdot A \cdot AaC \cdot CF & R_8 &= \frac{V_8 \cdot BC}{K_8 + BC} \\
R_4 &= \frac{V_4 \cdot AC}{2(K_4 + AC)} & R_9 &= \alpha_9 \cdot BC \cdot Ah \\
R_5 &= \alpha_5 \cdot AC \cdot Ah & R_{10} &= \frac{V_{10} \cdot AaC}{K_{10} + AaC}
\end{aligned} \tag{5.16}$$

Here, Michaelis–Menten-like expressions, e.g. R_1, describe enzymatic conversions for which we assume a constant enzyme concentration. For pH-induced enzymes, we cannot apply the approximation of a constant total enzyme concentration in the derivation of the rate law. Hence, we obtain a multiplicative rate law, e.g. R_3, assuming a quasi-stationary intermediary complex[53]. Then, three additional differential equations present the time-dependent amount of the enzymes Adc, CtfA/B, and AdhE:

$$\begin{aligned}
\frac{dAd}{dt} &= r_{Ad} + r_{Ad}^{+} \cdot F - D \cdot Ad \\
\frac{dCf}{dt} &= r_{Cf} + r_{Cf}^{+} \cdot F - D \cdot Cf \\
\frac{dAh}{dt} &= r_{Ah} + r_{Ah}^{+} \cdot F - D \cdot Ah
\end{aligned} \tag{5.17}$$

where r_i denotes a basal level of protein synthesis during the acidogenic phase and r_i^{+} is the increment of the synthesis rate during solventogenesis. We introduced the step-like function

$$F(pH) = \frac{c_1}{2}\left[1 - \tanh\left(\frac{c_2}{2}\left(pH - c_3\right)\right)\right] + c_4 \tag{5.18}$$

to express the pH-dependent transition between both rates during the metabolic shift. Its parameters c_1 and c_4 are determined by the initial and the final pH value. The further constants are estimated from the time course of the external pH. Furthermore, we consider a dilution term to account for the experimental set-up.

Several dynamic shift experiments have been conducted by our experimental collaborators providing data for parameter estimation and comparison of model and cellular behavior. These dynamic shift experiments[52] started at pH 5.7 for approximately 130 hours. Within this time, the clostridial culture established an acidogenic steady state. During a subsequent period of ~30 hours, without an external regulation of the pH level, the metabolic shift was initiated. Eventually, the clostridial culture approached a solventogenic steady state at pH 4.5. The solventogenic steady state was maintained for more than 200 hours until the end of the experiment. During these experiments, the concentrations of metabolites acetate, butyrate, ethanol, acetone, and butanol had been measured. Furthermore, the external pH was determined.

Using these experimental data, we estimated the kinetic parameters of Eq. (5.16) and the synthesis rates of Eq. (5.17)[52]. Afterwards, we simulated numerically the system of coupled differential equations. In Fig. 5.10, we compare our simulation (solid lines) with the time courses of the products measured in the experiments (dots). In agreement with experimental data, the simulation establishes an acidogenic steady state at which the acid production dominates. Later, the external pH level changes triggering the metabolic shift. This results in a fast decrease of the acid concentrations. Simultaneously, the solvent-producing enzymes are induced. Consequently, the metabolic output changes to solvent production. Finally, the simulation approaches a solventogenic steady state. In accordance with experiments, the main fermentation products during this phase are the solvents acetone and butanol.

Our model describes the pH-induced metabolic shift in *C. acetobutylicum* as a multilevel cellular process involving alterations

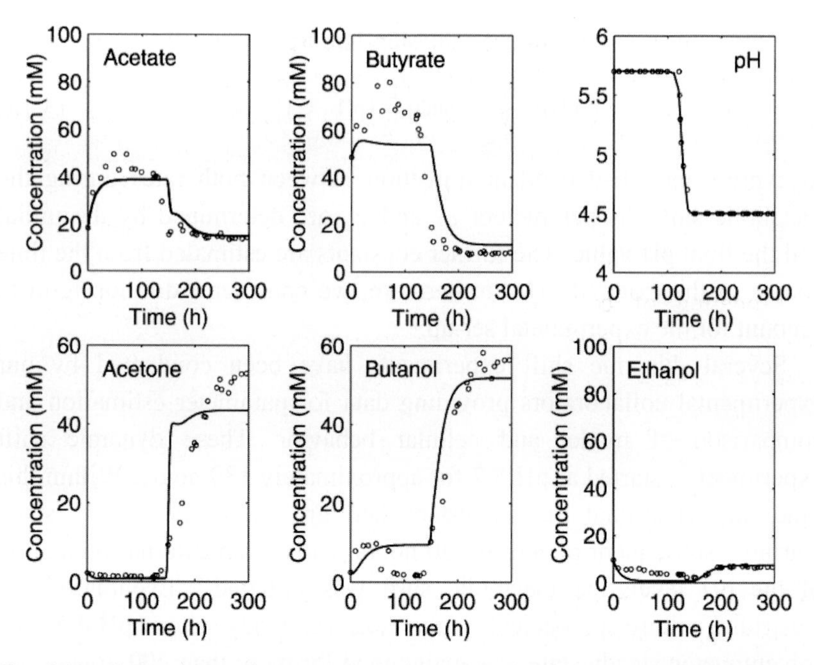

Fig. 5.10: Comparison of a numerical simulation (solid lines) of Eq. (5.16) and experimental data (dots) of a dynamic shift experiment[52]. In agreement with the experimental data, the simulation exhibits a pH-dependent metabolic switch from acid to solvent production. The time courses of the acids acetate, butyrate, and the pH are plotted in the upper row. The solvents acetone, butanol, and ethanol are shown in the lower row.

on transcriptomic, proteomic, metabolomic, and environmental levels. It successfully demonstrates that their temporal changes and interrelations have to be considered for the description of the pH-induced metabolic shift.

5.4 Conclusions and outlook

Dwindling fossil fuel resources and climate changes caused by their use challenge future economic and ecologic systems. To overcome these challenges, new technologies and strategies for industrial production and use of related chemicals are required. Biotechnological approaches seem to be a promising alternative to past and current methods. However, they have to fulfill some criteria arising from industrial, political, and safety demands. Furthermore, future industrial application requires an improvement of efficiency of the biological processes. Hence, a sophisticated knowledge about the biological systems, such as the AB fermentation in *C. acetobutylicum*, is fundamental for a successful large-scale industrial application. It could provide information for the design of optimal industrial facilities and of optimized microbial strains that fit the industrial objectives.

Modeling of cellular behavior is a prerequisite for large-scale industrial application. It provides information about the underlying biochemical mechanisms, involved cellular levels, regulations, and adaptations. Biological networks, like AB fermentation in *C. acetobutylicum*, are characterized by high complexity and non-linearity which require mathematical approaches to understand the behavior. Here, we used the pH-induced metabolic shift as an example in order to demonstrate how biological information is used to establish a mathematical model. Our model successfully reveals that this phenomenon includes several levels of cellular organization. However, the relationship between these levels is complex and partially unknown. Thus, mechanistic models of their dynamic interactions have to be established in future.

In a first step, we established the framework of open systems which reflects the chosen experimental set-up, a pH-controlled continuous culture. This illustrates that a model of cellular systems has to consider the environment the cells are living in. In turn, this modeling requirement also creates demands on the experiments. Modeling depends upon reproducible biological data. The standardized experimental set-up was chosen for that purpose.

In agreement with recent experimental data, we demonstrated that changes in transcriptomic and proteomic profiles are not necessarily directly proportional either in their dynamic or their steady-state behavior. These findings suggest that a multilevel perspective in modeling and experiment is required to explain the dynamic adaptation of cellular systems. In particular, the cellular response to changes in physical parameters, such as pH and temperature, which impact kinetic properties of the processes involved, necessitates such an approach.

Emerging experimental evidence indicates that a remarkable rearrangement of the cellular composition accompanies the pH-induced metabolic switch. This suggests that the bacterium senses changes in the extra- and intracellular pH level. However, the underlying mechanisms are still unknown. We suggest that a protein or transcription factor that exhibits a pH-dependent activation status acts as a pH sensor. This sensory element triggers the transcriptional changes observed in recent experiments.

Our model of the dynamic pH-shift demonstrates that this phenomenon can be explained only if adaptations on transcriptomic, proteomic, and metabolomic levels are considered sufficiently. Hence, a sophisticated knowledge of these pH-dependent cellular levels and their interrelationships is required to improve future modeling approaches. Their predictions could provide the basis for purposeful mutations of *C. acetobutylicum* regarding increased solvent production. The present model was able to show that a single mutation is insufficient for that purpose.

In summary, the modeling of the pH-induced metabolic shift in *C. acetobutylicum* merges dynamic information of several levels of biological organization into a joint representation. By this means, it provides a unique way of investigating the complex network of AB fermentation. Futhermore, new questions arise from the predictions made by simulations, missing information, or different model structures reproducing the observed behavior. Additionally, model-driven design of experiments could help to unravel the complex networks of cellular behavior.

Acknowledgement

The authors acknowledge the support of the German Federal Ministry for Education and Research (BMBF) (FKZ 0313981D, 0315782D, 0313978F, and 0315784E) as part of the European Transnational Network Systems Biology of Microorganism (SysMO) within the COSMIC and BaCell consortia.

References

1. Rittmann B. E., *Biotechnol. Bioeng.* 100 (2008).
2. International Energy Agency (IEA), *Key World Energy Statistics 2011* (2011).
3. Dürre P., *Biotechnol. J.* 2 (2007).
4. Jones D. T. and Woods D. R., *Microbiol. Rev.* 50 (1986).
5. Bahl H. *et al., Appl. Microbiol. Biotechnol.* 14 (1982).
6. Bahl H. *et al., Appl. Microbiol. Biotechnol.* 15 (1982).
7. Gottwald M. and Gottschalk G., *Arch. Microbiol.* 143 (1985).
8. Janssen H. *et al., Appl. Microbiol. Biotechnol.* 87 (2010).
9. Millat T. *et al., Microbial Biotechnol.* 6 (2013).
10. Janssen H., Ph.D. Thesis, University of Rostock, Germany (2010).
11. Fischer R. J. *et al., J. Bacteriol.* 188 (2006).
12. Tangney M. and Mitchell W., *Appl. Microbiol. Biotechnol.* 74 (2007).
13. Chen J. S., *FEMS Microbiol. Rev.* 17 (1995).
14. Dürre P. *et al., FEMS Microbiol. Rev.* 17 (1995).
15. Andersch W. *et al., Appl. Microbiol. Biotechnol.* 18 (1983).
16. Hartmanis M. G. N. *et al., Appl. Microbiol. Biotechnol.* 20 (1984).
17. Fischer R. J. *et al., J. Bacteriol.* 175 (1993).
18. Walter K. A. *et al., J. Bacteriol.* 174 (1992).
19. Grimmler C. *et al., J. Mol. Microbiol. Biotechnol.* 20 (2011).
20. Williamson W. P., *Biochem. Soc. Trans.* 33 (2005).
21. Schwarz K., *et al., J. Microbiol. Meth.* (2007).
22. Gray P. and Scott S., *Chemical Oscillations and Instabilities: Nonlinear Chemical Kinetics*. (Oxford University Press, Oxford, 1994).
23. Keene J. D., *BMC Biol.* 8 (2010).
24. Alberts B. *et al., Molecular Biology of the Cell*, 4th Edition (Garland Science, New York, 2002).
25. Madigan M. T. *et al., Brock Biology of Microorganisms*, 12th Edition (Prentice Hall International, New Jersey, 2009).
26. de Sousa Abreu R. *et al., Mol. BioSyst.* 5 (2009).
27. Nie L. *et al., Bioinformatics* 22 (2006).
28. Nie L. *et al., Genetics* 174 (2006).
29. Batschelet E. *Introduction to Mathematics for Life Scientists*, 3rd Edition (Springer-Verlag, Berlin, 1979).
30. Murray J. D. *Mathematical Biology I: An Introduction*, 3rd Edition (Springer-Verlag, Berlin, 2002).
31. Kühner S. *et al., Science* 326 (2009).
32. Güell M. *et al., Science* 326:1268–1271 (2009).
33. Sachleben J. R. *et al., P. Natl. Acad. Sci. USA* 107 (2010).
34. Shen J. K. *Biophys. J.* 99 (2010).
35. Segel I. H. *Enzyme Kinetics* (John Wiley and Sons, New Jersey, 1993).

36. Alberty R. A. and Massey V., *Biochim. Biophys. Acta* 13 (1954).
37. Heinemann M. and Sauer U., *Curr. Opin. Microbiol.* 13 (2010).
38. Atkins P. W. and de Paula J., *Atkins' Physical Chemistry*, 7th Edition (Oxford University Press, Oxford, 2002).
39. Pilling M. J. and Seakins P. W., *Reaction Kinetics* (Oxford University Press, Oxford, 2001).
40. Cornish-Bowden A., *Fundamentals of Enzyme Kinetics*, 3rd Edition (Portland Press, London, 2004).
41. Bisswanger H., *Enzyme Kinetics: Principles and Methods* (Wiley-VCH, Weinheim, 2002).
42. Fowler R. and Guggenheim E. A., *Statistical Thermodynamics* (Cambridge University Press, Cambridge, 1952).
43. Michaelis L. and Davidsohn H., *Biochem. Z.* 35 (1911).
44. Michaelis L. *Die Wasserstoffionenkonzentration* (Springer, Berlin, 1922).
45. Flach E. H. and Schnell S., *IEE Proc. Syst. Biol.* 153(2006).
46. Millat T. *et al., Biosystems* 92 (2008).
47. Garcia-Moreno B., *J. Biol.* 8 (2009).
48. Booth I. R., *Microbiol. Rev.* 49 (1985).
49. Huang L. *et al., Appl. Environ. Microbiol.* 50 (1985).
50. Nakamura S. *et al., J. Mol. Biol.* 386 (2009).
51. Yu X. J. *et al., Science* 328 (2010).
52. Haus S. *et al., BMC Syst. Biol.* 5 (2011).
53. Millat T. *et al., Math. Biosci.* 207 (2007).
54. McNaught A. D and Wilkinson A., *Compendium of Chemical Terminology*, 2nd Edition (Blackwell Science, New Jersey, 1997).

CHAPTER 6

MATHEMATICAL MODELS FOR CLOSTRIDIA: FROM CULTIVATION DESCRIPTION TO SYSTEMS BIOLOGY

S. JUNNE

Technische Universität Berlin
Department of Biotechnology, Chair of Bioprocess Engineering
Ackerstrasse 71-76, 13355 Berlin, Germany

P. GÖTZ

Bioprocess Engineering
Beuth University of Applied Sciences Berlin
Seestrasse 64, 13347 Berlin, Germany

In an industrial surrounding, process models are useful for process evaluation, prediction, and control. The abstraction of a process time course into an empirical equation may be sufficient under these conditions, yet this type of model cannot be extrapolated and does not provide further insight into the process. The development of mechanistic process models, based on transport phenomena and the corresponding balance equations, yields an improved process description and additionally a set of interpretable model parameters. Extending this concept to its limits, using the availability of molecular data on various levels of intracellular activities, may lead to the ultimate mathematical model of a cell interacting with its environment. This model will reflect the actions of molecules within the cell, creating the complex systems behavior that is observed in cellular processes. It will accommodate all data created when working with cells and allow prediction of events within the cell in interrelation with its environment. From a long history of interest in and applications of clostridia, model development now evolves from simple descriptions towards systems biology and the *in silico Clostridium*.

6.1 Introduction

There has always been an interaction between clostridia and humans, ranging from the pathogenic effects of toxins in food spoilage and infection to the industrial production of solvents. Therefore there should be considerable interest in predicting and controlling the behavior of this class of microorganisms. In an industrial environment, prediction of the time course of growth and metabolite production is usually accomplished through the development of a mathematical model. Such process models, as they are understood in this chapter, are able to reflect growth, substrate consumption, and product synthesis based on empirical and/or kinetic equations. Mathematical models for clostridia may also provide a basis for risk assessment of contamination in food treatment, where the prediction of bacterial growth is of great importance. Currently, the main application of clostridia in bioprocesses for solvent production is challenging due to the complexity of the regulation of growth and product synthesis, and the sensitivity to alterations in the environome. These features lead to very complex responses of the microorganism. The representation of this complexity in a mechanistic process model may provide support in coping with these issues. Changes in product composition can be predicted, and process parameters can be adapted based on these predictions. Ideally, a fast and targeted response by model-based process control to changes in the bacterial behavior becomes possible. In this case, models can be used to reduce process costs (e.g. by optimization of feeding and *in situ* removal strategies), to enhance the stability of the process when combined with a powerful control unit, and to speed up strain development for industrial applications. Process models are also suitable as a basis for a further systems biology approach for the understanding of the cell's behavior on a molecular level. This approach extends the analysis to the molecular composition of the cell (genome, transcriptome, proteome, and metabolome) and related dynamic changes (fluxome) including the processing of information within regulatory networks. Mathematical models for the physiology and metabolism of the cell can be derived from this approach. Starting with mass balances for the substrates, intermediate metabolites, and products, these equations are linked by respective fluxes (transport and reaction), which in turn are influenced by protein levels being controlled by transcription of genomic information. Additional influence is exerted by regulation, which can be represented, for example, by an independent class of regulatory molecules that interact with DNA.

In order to characterize and classify the multitude of different approaches to model cell cultivation-based bioprocesses, an overview of the different types is presented in Fig. 6.1.

The general classification of model types in Fig. 6.1 differentiates between the macroscopic and the molecular level. The macroscopic level ranges from a simple yield calculation (Type 1) to simple structured models (Type 3) based on mass balances. Mass balance-based models are able to predict process performance and are the prevailing mathematical tool in the bioprocess industry.

On a molecular level, the model development is based on a metabolic network, which comprises all pathways that may be present in the organism. Usually, this network is derived from the genome sequence by functional annotation of the genes. From the static information in the network, reaction routes from substrate to product can be identified and theoretical yields can be predicted from the stoichiometry of these reactions (Type 4 and 5). Furthermore, metabolome data can be checked for consistency by applying the constraints derived from these relations.

Evaluation of metabolome data for gaining insight into intracellular processes is accomplished by calculating the metabolite fluxes within the cell at (quasi-) steady state (Type 6). Depending on the complexity of the (sub-)network in view, a large number of metabolites has to be measured. The result is the description of fluxes between the metabolite pools in the cell. Split ratios at branching points in the network are of special interest, they enable the identification of possible bottlenecks to be removed by metabolic engineering. Type 6 is only a descriptive method, that is it is valuable for comparison of different strains, but it can not predict, for example, the outcome of a genetic modification.

The transition from evaluation of (quasi-)stationary molecular data for gaining a snapshot of the physiological status of the cell to dynamic time series data increases the analytical effort. Following the changes of the metabolome over time and calculating the corresponding dynamics of the fluxes between the metabolite pools (which are the enzymatic reaction rates of the biochemical conversions between these pools), dynamic experiments (e.g. pulse experiments) are evaluated and the

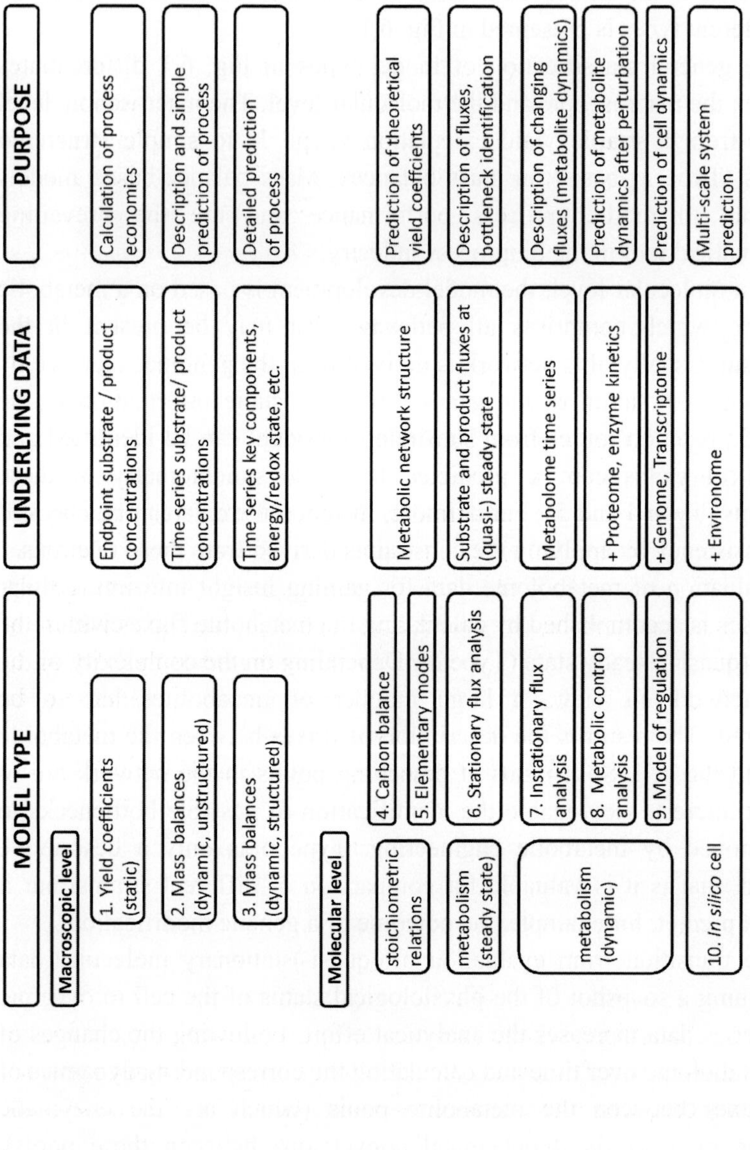

MODEL TYPE		UNDERLYING DATA	PURPOSE
Macroscopic level			
	1. Yield coefficients (static)	Endpoint substrate / product concentrations	Calculation of process economics
	2. Mass balances (dynamic, unstructured)	Time series substrate/ product concentrations	Description and simple prediction of process
	3. Mass balances (dynamic, structured)	Time series key components, energy/redox state, etc.	Detailed prediction of process
Molecular level			
stoichiometric relations	4. Carbon balance 5. Elementary modes	Metabolic network structure	Prediction of theoretical yield coeffcients
metabolism (steady state)	6. Stationary flux analysis	Substrate and product fluxes at (quasi-) steady state	Description of fluxes, bottleneck identification
metabolism (dynamic)	7. Instationary flux analysis	Metabolome time series	Description of changing fluxes (metabolite dynamics)
	8. Metabolic control analysis	+ Proteome, enzyme kinetics	Prediction of metabolite dynamics after perturbation
	9. Model of regulation	+ Genome, Transcriptome	Prediction of cell dynamics
10. *In silico* cell		+ Environome	Multi-scale system prediction

Fig. 6.1: Types of models.

impact of the experimental conditions is quantified (Type 7). From the dynamics of the fluxes, hypotheses on inhibition, activation, induction, and/or repression can be derived, but since this model type is also only descriptive, a verification of the hypotheses by experimental design prediction is not possible.

Adding proteome time series data to the metabolome time series data leads to the modeling approach, Type 8. By separating protein level (enzyme concentration) and enzyme activity (kinetic behavior), the reaction rates between the metabolite pools can be predicted from proteome and metabolome data. The mathematical formulation of kinetic behavior is usually based on a Michaelis–Menten-type equation, expanded by inhibition, activation, allosteric, or other mechanisms. The knowledge of the kinetic behavior enables the use of metabolic control analysis, which yields information on the effect of perturbations of metabolite concentrations within the network. If proteome dynamics play a minor role in the process, then a detailed prediction of metabolite dynamics in the process is possible.

Batch, shift, or pulse experiments force dynamic changes of the environment upon the cells, resulting in an adaptation beyond enzyme kinetic behavior. Signal transduction and regulatory processes govern this adaptation, leading to interaction between various classes of molecules, from activation of signal proteins by phosphorylation to induction of DNA transcription and translation. A genome-wide metabolic network (the basis for Type 4 and Type 5) is not always evenly active, but depends on regulation. The topology of the active network changes over time and different sub-networks arise. By monitoring the transcriptome, the sub-network topology can be tracked over time. Adding genomic information (operons, binding sites for induction, repression etc.), the interactions between all the players in regulation are mathematically described and model parameters are estimated to allow prediction of cell dynamics (Type 9).

The final goal, the *in silico* cell, will comprise prediction of molecular process dynamics within the cell coupled to the environmental conditions around the cell, which can also be modeled (Type 10). This multi-scale model, linking molecular events to the function of the microscopic cell in a macroscopic environment enables a detailed and accurate process description, inferring a deep understanding of process dynamics and allowing superior process prediction and process control.

6.2 Models for the ABE fermentation process with clostridia

6.2.1 History

The major industrial application of clostridia in biotechnological processes in the 20th century was the acetone-butanol-ethanol (ABE) fermentation[1]. The production of acids and solvents with *Clostridium acetobutylicum* and some other closely related *Clostridium* strains was one of the largest bioprocesses, second only to bioethanol production, during the first half of the 20th century[2]. Modeling the ABE fermentation is challenging, since it is divided into two main process phases. An acidogenic growth phase with the major products acetic and butyric acid is followed by a solventogenic phase, which is characterized by the production of the solvents acetone, butanol, and ethanol. Sporulation is initiated closely to the point at which the culture switches from the acidogenic to the solventogenic phase. Although this metabolic switch has been the focus of research for decades, the molecular mechanism behind it is still not fully understood. Hence, the development of a mechanistic model, which accounts for this event, remains a challenge. Despite the great importance of the ABE fermentation in former days, not many attempts for developing advanced process models can be found in the literature. However, it can be regarded as first step towards the quantitative description of the ABE fermentation, when systematic examinations of product ratios for utilization of different carbon sources[3,4] or media additives[5] were performed.

Early models for solvent bioproduction are reported for ethanol synthesis by yeast in batch[6-8] and in continuous processes[9-11]. The inhibitory effect of ethanol on the growth of yeast and on ethanol production was described by model equations under ethanol-producing conditions[12]. An inhibition ansatz was introduced for growth:

$$\mu=\mu_0 e^{-k1cEtOH} \tag{6.1}$$

and for the production of ethanol:

$$v = v_0 e^{-k2cEtOH} \tag{6.2}$$

where μ_0 and v_0 are values of the growth rate μ and ethanol production rate v when ethanol is not present, and k_1 and k_2 are model parameters. Although such an approach should have been possible for describing the ABE fermentation, there is no related report. Based on the pioneering work conducted on fermentation kinetics[8,13], a general methodology for classifying fermentation processes based on its kinetics was presented[14]. The growth rate, as well as the product formation rate, was expressed in separate equations in the exponential growth phase, in a constant growth phase, in the retardation phase characterized by slowing of growth, and a declining production phase (independent from the state of growth). By dividing the fermentation into these phases, growth and product formation was described for the examples of gluconic acid formation by *Pseudomonas ovalis* ATCC 8209 and of 5-ketogluconic acid fermentation by *Acetobacter suboxydans*. In these early days of bioprocess modeling, ABE fermentation was not yet the subject of detailed quantification.

6.2.2 The ABE fermentation with *C. acetobutylicum* in batch processes

One of the major platforms for the development of process models for the description of the ABE fermentation process was established by fermentation equations. They were used to describe the product yield and biomass formation based on sugar consumption with respect to known biochemistry and thermodynamic constraints[15,16]. The first report describing a fermentation equation for the ABE process[17] presented equations that described the adenosine triphosphate (ATP) utilization, the reduction energy demands including a balance of NADH, and hydrogen production with respect to glucose consumption, main product formation and growth. This resulted in four general equations containing 11 unknown factors representing the stoichiometry within the metabolism, when the concentration of non-accumulating intermediates was assumed to be zero. Inhibitory effects of the products led to a detailed study of kinetics in ABE fermentation[18]. Acetic and butyric acid, acetone, butanol, and ethanol were investigated separately with respect to their potential for inhibiting the growth of *C. acetobutylicum*. Growth rate under inhibition was expressed by the following equation:

$$\mu = \frac{\mu^{i}_{max}\, c_S}{K_S + c_S} \tag{6.3}$$

where μ^{i}_{max} represents the maximum growth rate under inhibition conditions. By definition of a function $F(I)<1$ which is related to the maximum growth rate without inhibition, the effect is quantified.

$$\frac{\mu^{i}_{max}}{\mu_{max}} = F(I) \tag{6.4}$$

From experiments, an empirical relationship for this function was found:

$$F(I) = 1 - \left(\frac{c_{inh}}{c_{max,inh}} \right)^{m_i} \tag{6.5}$$

where $c_{max,inh}$ represents the concentration of inhibitor *inh*, above which cell growth is completely inhibited. Experimentally determined maximum concentrations that caused growth inhibition were determined to be 12 and 11 g L^{-1} for acetic and butyric acid, and 17 g L^{-1} for butanol respectively. Combined inhibitory effects were modeled by extending Eq. (6.5) with additional expressions for *n* species of inhibitory compounds *j*. Including the proton concentration via pH, the time course of cell density could be predicted well in batch cultures at different pH values between pH = 4.1 and 4.5.

While the number of publications on kinetic studies and mathematical model development increased in the 1970s and 1980s, more complex process models were also derived for ABE fermentation. Among the first approaches is a predictive model, where butanol end-product inhibitory effects are included in the model as well as the RNA content of the cell. The differential equations are based on mass balances, the respective reaction rates primarily described by Monod-type kinetic equations[19]. Culture physiology is included in the model via an RNA-dependent state marker *y* to divide the two production phases:

$$\frac{d(y \cdot c_x)}{dt} = \mu(c_S, c_{BUOH}) \cdot y \cdot c_X \tag{6.6}$$

where *y* is defined as the dimensionless RNA concentration:

$$y = \frac{c_{RNA}}{c_{RNA,min}} \tag{6.7}$$

and $c_{RNA,min}$ is the RNA concentration in the cell at $\mu = 0$. The initial value for *y* for the solution of Eq. (6.6) is dependent on the state of the inoculum. If the inoculum is in the growth phase, $y(t = 0)$ is larger than 1. After an extended stationary phase (in the absence of sugar substrate), when not all cells are capable of growth, *y* may even be smaller than 1. In Eq. (6.6), the function $\mu(c_S, c_{BUOH})$ describes the growth depending on the substrate concentration c_S and the butanol concentration c_{BUOH}. The delay in solvent production is accounted for by introducing the conversion rates of acetic acid and butyric acid into the differential equations for solvent synthesis by a stoichiometric factor. Examination of the model equations shows that the authors do not accurately separate production, consumption, and accumulation of acids in their model derivation. Their specific rate q_{BA} is actually the accumulation of butyric acid, hence the sum of production from sugar and conversion into butanol. In their Eq. (6.13) for the butanol concentration, only the latter expression, multiplied by a stoichiometric factor, should be used for butanol production from butyric acid. By using dBA/dt instead, the authors imply that formation of butyric acid from sugar simultaneously causes butanol consumption. From a mechanistic viewpoint this is not correct, although using this mathematical formulation in simulations results in a pronounced delay of solvent synthesis.

The hypothesis that sugar transport through the cellular membrane is the major controlling step for acid and solvent production was the basis for another model of *C. acetobutylicum* ATCC 824 cultivation[20]. It was postulated that the cellular sugar transport (regarded as component A) is performed by active sites in the membrane (regarded as components X). The total concentration of the active membrane sites c_T is the sum of sites with bound substrate c_A and vacant sites c_X. This derivation inevitably leads to a Michaelis–Menten-type expression for the dependence of the sugar uptake rate from the sugar concentration by assuming n active binding sites:

$$q_S = n \cdot \frac{K_1 \cdot c_S}{c_S + K_S} \tag{6.8}$$

Introducing $n' = nK_1$ into Eq. (6.8), a change of the maximum sugar uptake rate n' is described by considering the physiologic state of the cells via introducing a marker y (dimensionless concentration of RNA)

$$\frac{dn'}{dt} = f(c_{BUOH,int}) \cdot y - n' \cdot \mu \tag{6.9}$$

where $f(c_{BUOH,int})$ is a function of the intracellular butanol concentration. It is estimated based on experimental data, when n' is constant. Similar to the sugar transport, also the butanol transport (from the inside to the outside) is assumed to be based on active sites in the membrane. Despite this hypothesis, the authors formulate the transport rate based on a driving concentration difference and a cell membrane permeability P, therefore implying passive transport:

$$\frac{dc_{BUOH}}{dt} = q_{BUOH} \cdot X = P \cdot V_B \cdot (c_{BUOH,int} - c_{BUOH}) \tag{6.10}$$

where V_B is the wet cell volume. For the description of intracellular butanol synthesis, expressions from the aforementioned publication[19] were used. A major problem in modeling ABE fermentation is the delay in solvent production. The model expansion by addition of cross-membrane transport adds a differential equation (Eq. (6.10)), mathematically introducing a time delay. This will result in an improvement of delay representation without a necessary mechanistic background.

A more recently developed kinetic model for the ABE fermentation process attempts to separate acidogenesis and solventogenesis by adding more detail regarding metabolism[21]. The model is based on 19 metabolite mass balances from glucose to the end-products acetone, ethanol, and butanol for the cultivation of *C. saccharoperbutylacetonicum* strain N1-4. The reaction kinetics are Michaelis–Menten type, extended by glucose substrate inhibition, activation by butyric acid, and inhibition by butanol when appropriate. Additionally, the reactions catalyzed by acetoacetyl-CoA transferase are described by a random bi-bi mechanism. The metabolic switch is assumed to be triggered when the substrate (that is, glucose) is depleted. This on–off mechanism is represented by the factor *F* (Eq. (6.11)) in the model. F is a multiplying factor within all kinetic rate equations involved in glycolysis and acidogenesis. Thereby, this leads to two metabolic sub-networks and during simulations the model toggles between these at low glucose concentrations.

$$\begin{aligned} c_S &> 1.00\text{mM} \rightarrow F = 1 \\ c_S &\leq 1.00\text{mM} \rightarrow F = 0 \end{aligned} \qquad (6.11)$$

The model was applied for initial glucose concentrations between 36.1 and 295 mM. The model showed a squared correlation of 0.901 after parameter estimation from time-dependent experimental concentration values. Hence, the model is able to simulate the time-dependent profiles of substrate and main products. The glucose concentration-based switch remains questionable, since there are many reports where solvent formation at considerably higher glucose concentrations was observed. Also the presence of high acid concentrations triggers solventogenesis while the glucose concentration has no influence.

Based on the fermentation equation[17] as described above, an instationary metabolic flux analysis (MFA) with non-linear constraints was developed[22]. The accumulation rate of each of the species that are part of the metabolic network is determined following Eq. (6.12). x_i is the rate of accumulation of species i, a_{ij} is the stoichiometric coefficient of species i in pathway j, and r_j is the flux through the pathway j:

$$x_i = \sum_j a_{ij} r_j \qquad (6.12)$$

The set of equations developed from this balance can be represented in matrix notation by Eq. (6.13), where *A* is a M x N matrix of stoichiometric coefficients, r is a N-dimensional flux vector and x is a M-dimensional species accumulation vector for a network composed of M species and N reaction pathways.

$$Ar = x \tag{6.13}$$

Species are divided into exchangeables, which accumulate outside the cell, non-accumulating intermediates, and flow components (e.g. co-factors).

The final matrix contains a singularity involving the pathways for acetic and butyric acid formation, and acetic and butyric acid uptake. These fluxes are related to each other by their involvement in the production of acetone. To resolve this singularity, an additional constraint is required. It is created based on *in vitro* description of the acetoacetyl-CoA transferase (CoAT)[23], the enzyme that catalyzes both acid back-reactions. Based on kinetic observations, the CoAT constraint relates the ratio of the two extracellular concentrations of butyric and acetic acid (c_{AA}, and c_{BA}) to the ratio of the two fluxes of acetic and butyric acid uptake/conversion (r_{BYUP}, and r_{ACUP}). This constraint (Eq. (6.14)) for the flux model is quantified by identification of an empirical factor.

$$\frac{r_{BYUP}}{r_{ACUP}} = 0.315\frac{c_{BA}}{c_{AA}} \tag{6.14}$$

Performing MFA for the set of metabolite pools, the fluxes are calculated by the minimization of the sum of weighted squared residuals between observed and calculated concentration values. Inclusion of the non-linear constraint (Eq. (6.14)) results in the minimization of the objective function (Eq. (6.15)). The first term represents a sum of weighted squared residuals with $\hat{r}$ as the desired pathway flux vector and with W as the weighting matrix. The second term represents a reformulation of Eq. (6.14). A lower limit vector, l, and an upper limit vector, u, are introduced, defined by the user. For each measured component, the weighting factor is the measurement standard error, for intermediates, the weighting factor is the largest magnitude flux in which the corresponding intermediate appears.

$$\left\| \frac{A\hat{r}}{W} - \frac{x}{W} \right\| + (r_{BYUP}c_{AA} - 0.315 r_{ACUP}c_{BA})^2 \qquad l \le \hat{r} \le u$$

(with 6.15)

By applying this MFA, the impact of genetic modifications on acid and solvent synthesis has been thoroughly studied[24-28].

The calculated fluxes to and from metabolite pools within the network are not neccessarily reflected by an accumulation within the related metabolite pool. A metabolite pool may decrease despite increased formation of this metabolite from a precursor, this effect being caused by an even larger consumption of this metabolite into another product.

Performing MFA using Eq. (6.15), the impact of acid and solvent addition on the fluxome of the main carbon metabolism at pH-controlled (pH = 5.0) batch fermentations was surveyed in a recent study[29]. In the middle of the acidogenic growth phase or at the onset of the solventogenic phase, acid or solvent was added, leading to increased but tolerable concentrations within the physiologic range.

In Table 6.1, the effects of pulses in the acidogenic phase are compared to standard conditions without pulse. As shown in the first column of Table 6.1, a concentration step of 50 mM of acetic acid led to an inversion of the acetic acid flux from formation to consumption during the first 10 minutes after the pulse. The values have to be compared to no inversion at non-stimulus standard fermentation conditions in the last column of Table 6.1. In the following time course (20 hours), acetic acid formation was re-established, being higher than under standard conditions. Butanol formation was significantly increased directly after the acetic acid pulse.

A pulse of butyric acid during acidogenic growth leading to a concentration step increase of 50 mM had no significant impact on the observed rates. Acetic acid, acetone, and butanol formation were slightly enhanced in the long-term observation period of 20 hours.

The acetone pulse in the acidogenic phase with a concentration step increase of 50 mM evoked the strongest effect on metabolic fluxes. Directly after the pulse, fluxes to acetic and butyric acid as well as to all major solvents were significantly increased. In the long-term observation period, a relevant change compared to standard conditions was only seen for the butyric acid flux. Butyric acid was not consumed via acetone formation, but apparently transformed back via butyryl-phosphate. This

Table 6.1: Average fluxes in mmol g^{-1} h^{-1} after pulse additions of acids and solvents and under standard conditions (no stimulus) in the acidogenic growth phase in pH 5.0 controlled *C. acetobutylicum* ATCC824 batch fermentations. Left values: evaluation after 10 minute time interval. Right values: evaluation after 20 hour time interval.

	Added stimulus component					Reference
	Acetic acid	Butyric acid	Acetone	Butanol	Ethanol	range (no stimulus)
$r_{Acetate}$	−10.9 / 2.4	1.9 / 2.2	24.7 / 0.7	6.3 / 0.6	5.3 / 0.7	0.9–2.5 / 0.4–1.9
$r_{Butyrate}$	8.5 / 0.6	4.2 / 0.7	23.3 /-0.2	9.3 / 0.6	7.7 / 0.1	2.3–9.8 / 0.1–0.6
$r_{Acetone}$	0.0 / 2.6	0.3 / 2.3	44.1 / 1.2	2.0 / 0.8	2.9 / 1.0	0.2–0.3 / 1.4–1.7
$r_{Butanol}$	4.5 / 2.5	1.8 / 5.4	5.1 / 3.3	10.8 / 1.4	4.1 / 2.7	0.7–0.9 / 2.9–5.2
$r_{Ethanol}$	0.0 / 0.3	0.2 / 0.3	0.7 / 0.2	2.9 / 0.1	0.6 / 0.3	0.1–0.2 / 0.1–0.3

may be caused by a product inhibition on acetone formation which can be observed at early growth stages of batch fermentations, also preventing the corresponding butyric acid uptake.

The effects from butanol addition during acidogenesis (concentration step increase of 80 mM) were similar to the acetone pulse, except no long-term reduction for the butyric acid formation flux was observed. When ethanol was added during acidogenesis (concentration step increase of 100 mM), only a weak enhancement of solvent formation fluxes was observed. Except for ethanol formation itself, all averaged fluxes were reduced during the following 20 hours of cultivation compared to standard conditions.

A different picture of the impacts of pulse additions on the fluxome was obtained, when the stimuli were performed during the solventogenic phase (Table 6.2). Acetic acid addition led to a strong increase in the butyric acid formation during the first 10 minutes after the pulse. Solvent production was reduced in both the short- and long-term observation.

Table 6.2: Average fluxes in mmol g^{-1} h^{-1} after pulse additions of acids and solvents and under standard conditions (no stimulus) in the solventogenic phase in pH 5.0 controlled *C. acetobutylicum* ATCC824 batch fermentations. Left values: evaluation after 10 minute time interval. Right values: evaluation after 20 hour time interval.

	Added stimulus component					Reference (no stimulus)
	Acetic acid	Butyric acid	Acetone	Butanol	Ethanol	
$r_{Acetate}$	0.9 / 0.3	–0.5 / 0.3	7.1 / 0.2	4.8 / –0.1	9.0 / 0.2	1.7 / 0.8
$r_{Butyrate}$	0.6 / 0.0	1.4 / –0.8	0.6 / 0.0	0.6 / 0.0	8.5 / 0.0	0.3 / 0.1
$r_{Acetone}$	1.1 / 0.5	0.8 / 1.2	9.1 / 0.2	5.1 / 0.6	10.5 / 0.2	2.1 / 1.0
$r_{Butanol}$	1.0 / 1.1	2.2 / 2.2	15.7 / 0.3	3.9 / 0.4	40.5 / 0.5	6.4 / 1.7
$r_{Ethanol}$	0.1 / 0.1	0.1 / 0.1	3.8 / 0.2	3.1 / 0.2	7.6 / 0.0	0.4 / 0.3

Butyric acid addition caused a redirection of the acetic acid flux to acetyl-phosphate, while butyric acid formation and its consumption via acetone formation were enhanced. In the following 20 hours, the butyric acid addition in the solventogenic phase led to the reduction of acid formation (in particular butyric acid formation) and to an enhanced conversion towards acetone and butanol. Among the solventogenic phase solvent pulse experiments, the ethanol pulse showed the greatest impact on the fluxome. Very similar results were obtained for the addition of acetone and butanol.

Since the yield of solvents, especially butanol, is of great importance for the industrial application of the ABE fermentation process, the maximum concentrations obtained in pulse experiments (and as reference in standard batch fermentations without pulse addition) are summarized in Tables 6.3 and 6.4. When acetic or butyric acid was added in the acidogenic growth phase, acetic acid addition yielded a larger amount of butanol. Lower amounts of butanol were produced when solvents were added in the acidogenic phase.

Table 6.3: Maximum concentrations (in mmol g^{-1}) after pulse additions of acids and solvents in the acidogenic growth phase and under standard conditions (no stimulus) in pH 5.0 controlled *C. acetobutylicum* ATCC824 batch fermentations.

	Added stimulus component					Reference
	Acetic acid	Butyric acid	Acetone	Butanol	Ethanol	(no stimulus)
Max. acetone conc.	82	43	52[1]	15	31	42
Max. butanol conc.	161	119	79	122[1]	51	99
Max. ethanol conc.	14	11	8	4	61[1]	10

[1] Concentration includes the added amount of solvents.

Table 6.4: Maximum concentrations (in mmol g^{-1}) after pulse additions of acids and solvents in the solventogenic phase and under standard conditions (no stimulus) in pH 5.0 controlled *C. acetobutylicum* ATCC824 batch fermentations.

	Added stimulus component					Reference
	Acetic acid	Butyric acid	Acetone	Butanol	Ethanol	(no stimulus)
Max. acetone conc.	36	58	75[1]	58	39	42
Max. butanol conc.	100	182	88	251[1]	100	99
Max. ethanol conc.	12	13	12	23	51[1]	10

[1] Concentration includes the added amount of solvents.

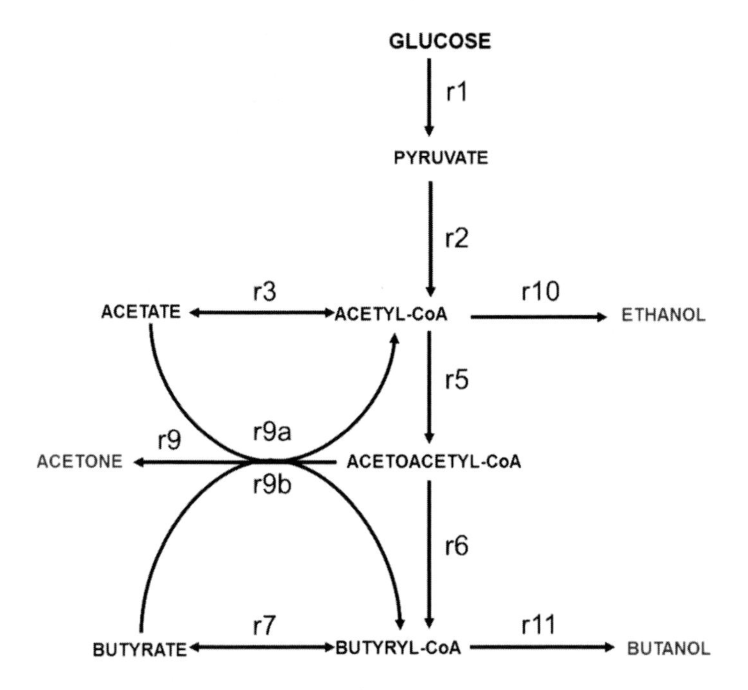

Fig. 6.2: Simplified metabolism for model development.

A different result was observed when acids and solvents were added in the solventogenic phase (Table 6.4). Butyric acid addition increased the amount of butanol produced, while butanol addition did not interfere with butanol production. The addition of acetone yielded lower amounts of butanol than under standard conditions.

From an initial kinetic model based on Michaelis–Menten-type equations, the observed interactions are now used to modify these equations. For an underlying model structure, a simplified metabolic network was constructed (Fig. 6.2).

The flux through pyruvate into acid and solvent formation was estimated including a yield coefficient *Y*, representing the part of the glucose uptake flux utilized for acid and solvent production. Balancing the pyruvate pool, Eq. (6.16) is derived:

$$\frac{dc_{\text{Pyruvate}}}{dt} = \frac{\rho_X}{c_X} * (Y * r1 - r2) - \mu c_{\text{Pyruvate}} \tag{6.16}$$

The term ρ_X / c_X is the inverse of the volume fraction of cells within the reactor, since extra- and intracellular compounds related to reactor and cellular volume respectively are both integrated in the model. All other balances are set up analogously. The corresponding rates contain the following extensions from Michaelis–Menten kinetics:

- Glucose uptake (r_1) is substrate and butanol inhibited
- Butyric acid formation from butyryl-CoA is product-inhibited and inhibited by butanol
- Acetic acid uptake coupled acetone formation is product (acetone) inhibited
- Butyric acid uptake coupled acetone formation is product (acetone) inhibited
- Ethanol formation is activated by acetic acid and inhibited by butyric acid and butanol
- Butanol formation is product inhibited

Beyond these enzyme kinetic considerations, a model of the metabolic switch has to take into account the changes in enzyme concentrations due to the regulation of enzyme expression. Experimental determination of transcriptome time courses[30] revealed a very large change of expression, in particular for the solvent-producing enzymes. An expression (Eq. (6.17)) is introduced, which is able to represent this change of expression by its sigmoidal shape. The induction of expression is formulated as a function of the extracellular butyric acid concentration. Butyric acid increase is generally considered to be one of the main causes for the onset of solventogenesis, this is supported by the results in Table 6.2.

$$\frac{c_{BA}^{\ n}}{c_{BA}^{\ n} + K^{\ n}} \tag{6.17}$$

Equation 6.17 yields values between 0 and 1. In combination with v_{max} this results in zero conversion for a complete lack of expressed enzymes and in a maximum rate at maximal enzyme expression. The rate equations for the model are summarized in Table 6.5.

Table 6.5: Rate equations for the network in Fig. 6.2.

$$r1 = c_X v_{\max 1} \frac{c_{Glucose}}{c_{Glucose} + K_{m1}(1 + \frac{c_{Glucose}}{K_{i,Glucose1}} + \frac{c_{BUOH}}{K_{i,BUOH1}})} \quad (6.18)$$

$$r2 = c_X v_{\max 2} \frac{c_{Pyruvate}}{c_{Pyruvate} + K_{m2}} \quad (6.19)$$

$$r3 = c_X v_{\max 3} \frac{c_{Acetyl-CoA}}{c_{Acetyl-CoA} + K_{m3}} \quad (6.20)$$

$$r5 = c_X v_{\max 5} \frac{c_{Acetyl-CoA}}{c_{Acetyl-CoA} + K_{m5}} \quad (6.21)$$

$$r6 = c_X v_{\max 6} \frac{c_{Acetoacetyl-CoA}}{c_{Acetoacetyl-CoA} + K_{m6}} \quad (6.22)$$

$$r7 = c_X v_{\max 7} \frac{c_{Butyryl-CoA}}{c_{Butyryl-CoA} + K_{m7}(1 + \frac{c_{BA}}{K_{i,Butyr7}} + \frac{c_{BUOH}}{K_{i,BUOH7}})} \quad (6.23)$$

$$r9a = c_X v_{\max 9a} \left(\frac{c_{BA}{}^n}{c_{BA}{}^n + K^n} \right) \frac{c_{AA}}{c_{AA} + K_{m9a} + \frac{c_{Acetone}}{K_{i,Acetone9a}}} \quad (6.24)$$

$$r9b = c_X v_{\max 9b} \left(\frac{c_{BA}{}^n}{c_{BA}{}^n + K^n} \right) \frac{c_{BA}}{c_{BA} + K_{m9b} + \frac{c_{Acetone}}{K_{i,Acetone9b}}} \quad (6.25)$$

$$r10 = c_X v_{\max 10} \left(\frac{c_{BA}{}^n}{c_{BA}{}^n + K^n} \right) \frac{c_{Acetyl-CoA}}{c_{Acetyl-CoA} + K_{m10} + \frac{K_{a,Acetate10}}{c_{AA}} + \frac{c_{BA}}{K_{i,Butyrate10}} + \frac{c_{BUOH}}{K_{i,BUOH10}}} \quad (6.26)$$

$$r11 = c_X v_{\max 11} \left(\frac{c_{BA}{}^n}{c_{BA}{}^n + K^n} \right) \frac{c_{Butyryl-CoA}}{c_{Butyryl-CoA} + K_{m11}\left(1 + \frac{c_{BuOH}}{K_{i,BUOH11}}\right)} \quad (6.27)$$

It is assumed that the intracellular concentration of acids reaches peak concentrations prior to the switch from acidogenesis to solventogenesis[31-33]. This can be justified by the low permeability of acids through the cell membrane. Therefore the intracellular acid concentrations were introduced into the model by additional differential Eqs (6.28) and (6.29) and the transport terms φ12 / φ13, for acetic acid.

$$\frac{dc_{AA,i}}{dt} = \left(\frac{\rho}{c_X} * (r3 - r9a - r12)\right) - \mu c_{AA,i} \tag{6.28}$$

$$\frac{dc_{AA}}{dt} = \varphi 12 \tag{6.29}$$

Accordingly for butyric acid:

$$\frac{dc_{BA,i}}{dt} = \left(\frac{\rho}{c_X} * (r7 - r9b - r13)\right) - \mu c_{BA,i} \tag{6.30}$$

$$\frac{dc_{BA}}{dt} = \varphi 13 \tag{6.31}$$

The transport of an acid A released (taken up) to (from) the medium by the cells depends only on the permeation of the undissociated and dissociated acid across the cell membrane:

$$\varphi_A = a_c (k_{un} (c_{AH_i} - c_{AH}) + k_d (c_{A^-i} - c_{A^-})) \tag{6.32}$$

If the Hendersson–Hasselbalch equation for the description of the ratio of concentrations of dissociated and undissociated acid concentrations based on the pH and pka value is applied, Eq. (6.32) can be rewritten as:

$$r_A = a_c \left(k_{un} \left(\frac{c_{A_i}}{1 + 10^{pHi - pka}} - c_{AH} \right) + k_d \left(\frac{c_{A_i}}{1 + 10^{pka - pHi}} - c_{A^-} \right) \right) * \frac{c_X}{\rho_X} \tag{6.33}$$

For acetic acid at pH = 5.0, the corresponding acid release can be determined:

$$\varphi_{12} = a_c \left(k_1 \left(\frac{c_{AA,un,i}}{18.38} - c_{AA,un} \right) + k_2 \left(\frac{c_{AA,d,i-}(t)}{1.06} - c_{AA,d} \right) \right) * \frac{c_X}{\rho_X} \tag{6.34}$$

Analogously for butyric acid:

$$\varphi_{13} = a_c (k_3 (\frac{c_{BA,un,i}}{16.14} - c_{BA,un}) + k_4 (\frac{c_{BA,d,i}}{1.07} - c_{BA,d})) * \frac{c_X}{\rho_X} \tag{6.35}$$

Model parameter estimation yielded a comparably low optimal cost (sum of squared errors between estimated and measured values) of 0.09. The optimal cost could be lowered to 0.06 by replacing the extracellular with the intracellular butyric acid concentration in Eq. (6.17) for enzyme expression. The final comparison between measured and simulated concentrations is shown in Fig. 6.3 and Fig. 6.4.

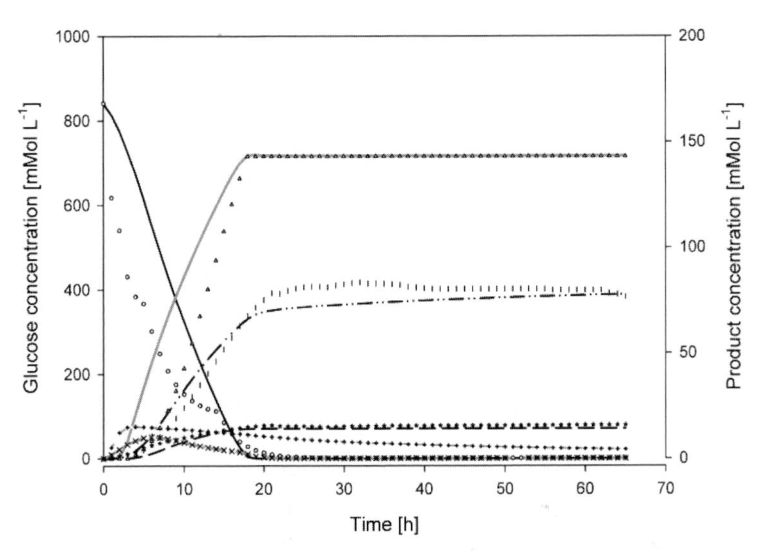

Fig. 6.3: Concentration profiles of substrate and products at a pH 5.0-controlled batch fermentation with *C. acetobutylicum* (dots) and the model simulation (lines): concentration development of glucose (open circle and black line), acetic acid (cross and dash-dotted gray line), butyric acid (filled plus and dotted gray line), butanol (open triangle and gray line), acetone (vertical mark and dash-dotted black line), and ethanol (filled star and dashed black line).

Major flaws of the model prediction of the measured substrate and product concentrations are:

- Glucose uptake is estimated too low at the beginning. This may be caused by the assumption of a constant Y
- The increase of butanol concentration is predicted too early.

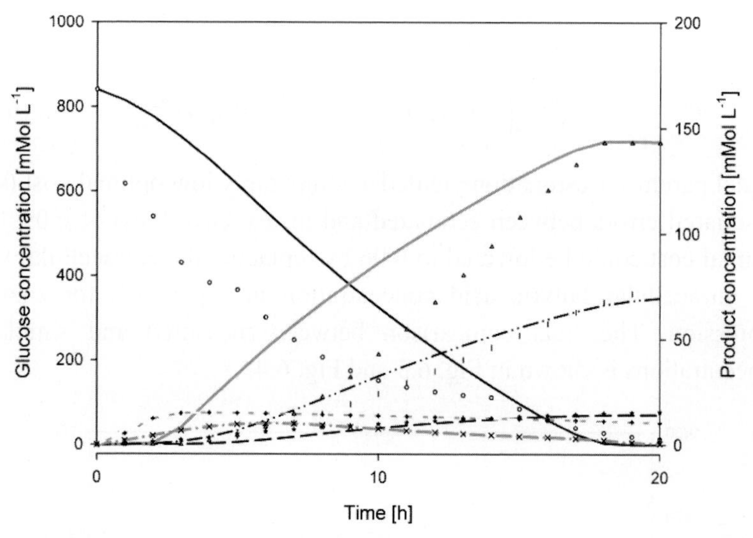

Fig. 6.4: Concentration profile of substrate and products at pH 5.0 in controlled batch fermentation with *C. acetobutylicum* (dots) and model simulation (lines) during the first 20 hours of cultivation: concentration development of glucose (open circle and black line), acetic acid (cross and dash-dotted gray line), butyric acid (filled plus and dotted gray line), butanol (open triangle and gray line), acetone (vertical mark and dash-dotted black line), and ethanol (filled star and dashed black line).

The model simulations for intracellular pH and intermediate concentrations are shown in Fig. 6.5a–6.5c. The intracellular pH is predicted to decline below 5.6 before the metabolic switch. The simulated intracellular concentrations of dissociated acetic and butyric acid concentrations reach peak values during this time. A more distinct development is seen for the intracellular dissociated acetic acid rather than butyric acid concentration development when the metabolic switch actually happens. The same is true for acetyl-CoA and butyryl-CoA, however, in both cases the simulated concentration development also accounts for the phosphorylated form, since it is not simulated separately due to the lack of measurement. Therefore, the simulation might include an accumulation of butyryl-phosphate prior to the onset of solventogenesis, as observed experimentally in a genetically engineered strain[34].

Table 6.6 provides an overview of the proposed mechanisms as they were applied in process models or assumed based on *in vitro*

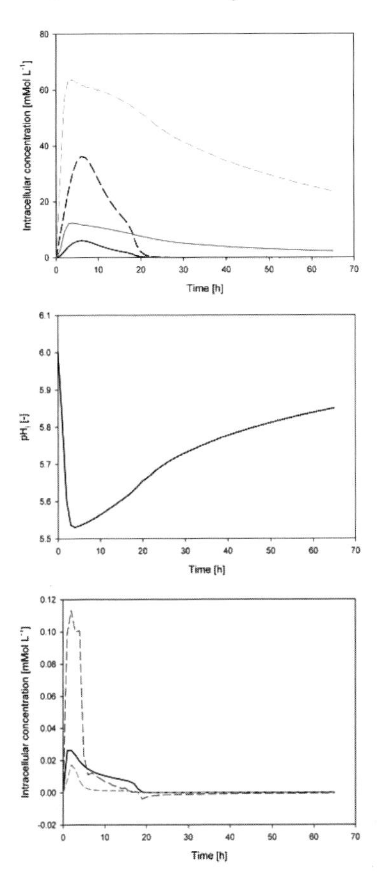

Fig. 6.5: Above — Simulation of intracellular acid concentrations in a pH 5.0 controlled *C. acetobutylicum* batch fermentation: concentration development of undissociated acetic acid (black line), dissociated acetic acid (dashed black line), undissociated butyric acid (gray line), and dissociated butyric acid (dashed gray line); Center – simulation of the intracellular pH in a pH 5.0 controlled *C. acetobutylicum* batch fermentation; Below – simulation of intracellular concentrations of intermediate metabolites in a pH 5.0 controlled *C. acetobutylicum* batch fermentation: concentration development of acetyl-CoA (black line), acetoacetyl-CoA (dashed black line), and butyryl-CoA (short-dashed gray line).

measurements and kinetic estimation of turnover rates for the corresponding enzymes.

The link between process models and models that consider regulation at various levels in the cell is approached by applying a proton flux

Table 6.6: Comparison of enzyme kinetic equations used in model development for *C. acetobutylicum* main carbon metabolism (equations that are extensions of standard Michaelis–Menten kinetics).

<table>
<tr><th>Enzyme</th><th colspan="3">Proposed mechanisms</th></tr>
<tr><td>Phosphotransferase
EC 2.7.1.69</td><td rowspan="3">Michaelis–Menten kinetic with substrate inhibition and uncompetitive inhibition by butanol[21]</td><td>Ping-pong mechanism in *B. subtilis* [35]</td><td rowspan="3">Michaelis–Menten kinetics with substrate inhibition and inhibition by butanol[29]</td></tr>
<tr><td>Phosphofructokinase
EC 2.7.1.11</td><td>Hill kinetics in *C.pasteurianum*[36]</td></tr>
<tr><td>Pyruvate-ferredoxin oxidoreductase
EC 1.2.7.1</td><td>Ping-pong-bi-bi mechanism[37]</td></tr>
<tr><td>Thiolase
EC 2.3.1.9</td><td>Ping-pong mechanism[23]</td><td>Multi-substrate ping-pong mechanism with product inhibition[38]</td><td></td></tr>
<tr><td>3-hydroxybutyryl-CoA dehydrogenase
EC 1.1.1.35</td><td>Mechanism with substrate and co-factor inhibition[36]</td><td></td><td></td></tr>
<tr><td>Crotonase
EC 4.2.1.17</td><td>Substrate inhibition[39]</td><td></td><td></td></tr>
<tr><td>Acetate kinase
EC 2.7.2.1</td><td>Hill kinetics[40]</td><td>Inhibition of butanol[19]</td><td></td></tr>
<tr><td>Acetaldehyde dehydrogenase
EC 1.2.1.10</td><td>Ping-pong mechanism[36]</td><td></td><td>Michaelis–Menten kinetics with inhibition by butanol and butyric acid[29]</td></tr>
<tr><td>CoA transferase
EC 2.8.3.8 &
EC 2.8.3.9:</td><td>Double competitive substrate inhibition[23, 38]</td><td>Random-bi-bi mechanism[21]</td><td></td></tr>
<tr><td>Butyrate kinase
EC 2.7.2.7</td><td>Ping-pong mechanism[36]</td><td></td><td>Michaelis–Menten kinetics with inhibition by butyric acid and butanol[29]</td></tr>
</table>

Table 6.6: (*Continued*)

Butanol dehydrogenase EC 1.1.1	Bi-bi mechanism with noncompetitive substrate and co-factor inhibition[36]	Uncompetitive inhibition of butanol and specific activation of butyric acid[21]	Michaelis–Menten kinetics with inhibition by butanol[29]

balance in a genome-scaled model for *C. acetobutylicum*[41,42]. In this model, the concept of specific flux proton states is introduced into constraint-based optimization of metabolic fluxes for the prediction of the extracellular medium pH. The specific proton flux state is thereby defined by a range of acceptable flux values across the cell membrane. The concept is based on the assumption that the rates of crossing the cell membrane by free protons can be used as an identifier for changes in the metabolic state of the cell. The extracellular proton concentration is assumed from a balance including the free protons calculated from pH measure-ments and the protons of major compounds of the cell suspension in the culture by a semi-mechanistic model. The specific proton flux through the membrane is determined based on the change of extracellular proton concentration. A genetic algorithm approach yields the requirement of six separate specific proton flux states to model sufficiently the acidogenic metabolism and to predict accurately the extracellular medium pH. Additionally, it could be demonstrated that the specific proton flux per weak acid decreases from 3.5 mol at the onset of batch fermentation to depletion at the end of the growth phase. It was found that proton flux through cation channels plays a role at the start of acidogenesis.

6.2.3 The ABE fermentation with *C. acetobutylicum* in continuous processes

Continuous cultivation of *C. acetobutylicum* in industrially applied ABE processes usually implies a continuous production process of several stages including the spatial separation of acid and solvent production. Such operation modes were realized at several production plants, e.g. in China as reported recently[43]. However, the development and application of process models for these processes are very rare. Due to the physiological changes which solvent-producing clostridia undergo between the acidogenic and solventogenic growth phase, and the connection to sporulation initiation, establishing a steady state in continuous culture is hardly possible. The prevailing way of performing a continuous cultivation is the implementation of either a limitation of a nutrient compound (e.g. sugar, nitrogen, phosphate) or an accumulation of an inhibitory product above a threshold concentration.

At the laboratory scale, several attempts have been made to achieve a continuous cultivation process with various strains of the genus *Clostridium*. Acid producers like *Clostridium butyricum* were successfully used in chemostat cultivations, since no switch from the acidogenic to the solventogenic phase occurs. It was found that the ratio of acetic to butyric acid produced depends on the dilution rate and culture pH. The ratio increased from 0.06 to 0.66 when the dilution rate varied between 0.02 and 0.29 h^{-1} at a pH of 6.0. An increased ratio was also observed when the glucose concentration or the pH level rose[44]. An optimum for hydrogen production during *C. butyricum* cultures was found at pH = 7.0 under phosphate limiting conditions at a dilution rate of 0.06 h^{-1} and a production rate of 2.22 mole per mole of consumed glucose[45]. A lower pH led to a higher incorporation of glucose into biomass. Hence, while the biomass yield increased, the hydrogen yield decreased. Another study of *C. butyricum* in a chemostat investigated the conversion of D-xylose[46]. While xylose availability was increased, the butyric acid production and biomass yield decreased. The hydrogen formation seemed to be rather independent of the substrate availability. When *C. butyricum* was cultivated in a continuous mode under carbon limited conditions using either glycerol or glucose as substrate, it was found that 1,3 propanediol formation remained stable when the dilution rate was increased. Lower specific synthesis of butyric acid resulted in a stronger accumulation of acetyl-CoA and NADH when applying glucose as substrate. In contrast, when using glycerol as substrate, the butyric acid formation was eight times higher at dilution rates of 0.08 h^{-1} [47]. The adverse ratio of $NADH/NAD^+$ can serve as a signal for a state of the cell where, due to regulatory mechanisms, bottlenecks in the metabolism led to lower carbon fluxes. This was also observed for *C. cellolyticum* cultivations when applying complex media. There the $NADH/NAD^+$ ratio was an order of magnitude higher than for cultivations with synthetic media. The production of hydrogen, ethanol, and lactic acid accounted for a balanced energy status of the cell. Optimal conditions for hydrogen production were found under ammonia limitation, leading to 2.08 mole per mole glucose consumed. Cellobiose carbon conversion in complex media led to accumulation of NADH and a reduced biomass yield[48]. This finding demonstrates the importance of the redox status of the cell. This parameter can serve as an indicator for limitations present in the cell during a production process. The pH dependence of *C. cellolyticum* chemostat cultivations was also observed, sporulation and substrate conversion reached highest values at pH = 7.0 at a dilution rate of $D = 0.053\ h^{-1}$ under glucose-limited conditions[49].

For cultivation processes yielding higher solvent concentrations, mainly two-stage approaches were described in the following publications[50–52]: in this operation mode, it is reported that solvent production reached values of $1.24\ gL^{-1}h^{-1}$ at a dilution rate of $0.13\ h^{-1}$ in *C. beijerinckii* cultivations. A maximum solvent concentration of $9.27\ gL^{-1}$ was achieved[50]. In another study, solvent concentrations of $0.15\ gL^{-1}$ with an overall productivity of $0.27\ gL^{-1}h^{-1}$ at dilution rates between $0.12\ h^{-1}$ in a first stage and $0.022\ h^{-1}$ in a second stage were achieved[51]. Interestingly, in these approaches, the carbon was not limiting growth but product inhibition. The integration of an evaporation loop increased the solvent yield by 40%[50]. Beyond solvent extraction, media extraction combined with a partial cell recovery can also reduce product inhibition. This operation mode led to a solvent productivity of $7.55\ gL^{-1}h^{-1}$ at a dilution rate of $0.11\ h^{-1}$ in *Clostridium saccharo-perbutylacetonicum* N1–4 cultivations. Maximum solvent concentrations of $8.58\ gL^{-1}$ were determined[52].

All the results summarized are generally suitable for the development of process models. The big advantage of a continuous culture is that the measurements of substrate, product, and metabolite concentrations can easily be converted into rates and fluxes when considering the corresponding dilution and growth rate. Hence, concentration data at steady state already imply fluxes without further mathematical evaluation. Some model approaches were developed beyond this simple stage by including kinetic considerations. The culture reduction state was determined by measuring the cellular NADH content at different dilution rates at a chemostat cultivation of *C. acetobutylicum* strain ATCC 824[53]. The specific substrate utilization rate was related to the specific growth rate as follows:

$$q_S = k_1\mu + k_2 \quad (6.36)$$

where k_1 reflects the sum of reciprocal biomass yields and k_2 reflects the sum of substrate consumption for maintenance and not growth-associated product formations. They were determined to be $k_1 = 8.5$ and $k_2 = 0.3$ at dilution rates 0.05 1/h $< D < 0.5$ 1/h. The following linear relationship between the NADH content of the culture and the butanol concentration was suggested:

$$c_{BUOH} = 0.05F - 2.98 \quad (6.37)$$

where F stands for the NADH fluorescence signal measured in the filtered fermentation broth. Keeping the fluorescence signal of the culture broth constant resulted in a controlled and stabilized butanol formation.

A kinetic approach was applied to develop a mathematical model for the continuous ABE fermentation process including cell retention[54]. The authors consider a state of low product concentration in the environment, when no (or not detectable) inhibition is present. In this state, growth follows the Monod equation. If growth inhibition by rising product concentration can not be neglected anymore, an additional product inhibition term is introduced:

$$\mu_i = \mu * f(I) \quad (6.38)$$

where *f(I)* represents the product inhibition model. It was defined as being case-dependent based on the overall butanol and butyric acid concentration (BBA). Two functions were applied, one for the case that BBA ≤ 8.0 gL^{-1}, one for the case $8.0 \leq$ BBA ≤ 13.9 gL^{-1}. The substrate uptake was expressed similarly to Eq. (6.36), including two extensions which consider that the conversion of acids to solvents requires energy that should be generated from substrate uptake. The conversion rate from butyric acid to butanol was therefore:

$$qBA = K_6 f(cBA)\, g(cS)\, cX \quad (6.39)$$

which is equivalent to the Luedeking–Piret model equation[55]. $f(c_{BA})$ represents the model equation describing the concentration of butyric acid and $g(c_S)$ the substrate concentration. The rates for the synthesis of acetic acid and its uptake were created similarly, as well as the butanol production. The model parameters were estimated by non-linear regression of a multiresponse mathematical model.

An increase of undissociated butyric acid concentration beyond a critical value is assumed to trigger solventogenesis. A concentration of 11.5 gL^{-1} of butyric acid irrespective to the extracellular pH was regarded as critical[56]. This hypothesis was introduced in a model for substrate limited chemostat cultivations[57]. It was further assumed that undissociated butyric acid passes freely through the membrane, hence the extra- and intracellular concentrations being equal. The model includes a product inhibition assumed for the butyric acid formation and static relations of acetic and butyric acid consumption and acetone synthesis. The inhibition term included in the model and described earlier for *C. thermocellum*[58] is:

$$I_{BA} = \frac{1}{1 + 4.5kc_{BA,un}(c_{BA,d,i} - c_{BA,d})} \tag{6.40}$$

where $c_{BA,un}$ represents the extracellular concentration of undissociated butyric acid, and $c_{BA,d}$ and $c_{BA,d,i}$ the extracellular and intracellular concentration of dissociated butyric acid, respectively. The conversion from butyric acid to butanol was assumed as a function of the gradient between the intra- and extracellular concentration of dissociated butyric acid:

$$r_{BA,uptake} = k\frac{c_S c_X}{0.2 + c_S}(4.5kc_{BA,un}(c_{BA,d,i} - c_{BA,d})) \tag{6.41}$$

The acetic acid uptake and acetone synthesis were assumed to correspond to the assumptions described in Eq. (6.41). Applying the resulting model, it was possible to reflect the differences in product spectra at acidic and neutral media pH in chemostat cultivations. So far, this model extension has not been applied to *C. acetobutylicum* batch fermentations.

Only recently, sampling and analysis procedures of chemostat cultivations were extended to allow for a systems biology approach. Phosphate-limited chemostat cultures controlled at a pH of 5.7 were operated at a steady-state acidogenesis. In the same cultures, at pH 4.5, the cells are transfered to solventogenesis without the initiation of sporulation. With the combination of state-of-the-art analysis on the transcriptomic, proteomic, and metabolomic level a systematic understanding of the acidogenic and solventogenic state of the cells and the underlying differences became possible[59]. The data gained in these experiments can provide models describing cellular processes beyond the main carbon metabolism on the metabolome level as e.g. gene regulation[60] and interactions related to enzyme expression and activity.

6.3 Models for processes with clostridia beyond the ABE fermentation

Bioenergetics and end-product regulation during *C. thermo-saccharolyticum* fermentations were the basis for corresponding model development[61]. Bacterial growth and substrate consumption were combined in a balance equation, considering the consumption of ATP by growth and maintenance and the ATP gain G_{ATP}. The gain represents the ratio of moles of ATP synthesized per moles of substrate consumed. Under consideration of the biomass yield coefficient $Y_{X/S}$, the following equation was developed:

$$q_S(1 - Y_{X/S}) * G_{ATP} = \frac{\mu}{Y_{ATP}^{max}} + m + w \tag{6.42}$$

where *m* represents the ATP consumed for the maintenance requirements and *w* the wastage due to loss of ATP. The maximum ATP-based cell yield Y_{ATP}^{max} was determined to be between 31 and 11.6 g cells/mol ATP, depending on the strain and carbon source. Different responses could be observed due to nutrient starvation with respect to G_{ATP} by the production of secondary metabolites.

Product inhibition of butyric acid in the fermentation of *C. butyricum* has been the subject of several reports. *C. butyricum* fermentations with respect to product inhibition in a pH-auxostat at low cell density and product concentration at dilution rates near the maximum growth rate were studied[62]. By the extracellular addition of products, their inhibitory effects were investigated. Concentrations of as high as 60 gL^{-1} of 1,3–propanediol, 27 gL^{-1} of acetic acid, and 19 gL^{-1} of butyric acid blocked growth at a culture pH of 6.5. A similar model as presented in Eq. (6.6) to simulate inhibitory effects was applied for modeling growth of *C. butyricum*[63]. As already reported in a previously mentioned publication[62], the inhibitory effects of acetic and butyric acid, propanediol, and glycerol were considered in Eq. (6.43):

$$\mu = \mu_{\max} \frac{c_S}{K_S + c_S} \left[\Pi \left(1 - \left(\frac{c_P}{c_P^m} \right)^n \right) \right] \tag{6.43}$$

Another approach is the use of a logistic function for the simulation of cell growth of *C. butyricum* for the production of butyric acid [64]:

$$\frac{dx}{dt} = \mu_{\max} \left(1 - \frac{c_x}{c_{x,\max}} \right) c_x \tag{6.44}$$

Product formation was simulated based on the Luedeking–Piret equation[55] (Eq. (6.39)). A corresponding equation was used for the simulation of the substrate uptake. It was assumed that the carbon utilized for product formation can be neglected in comparison to the amount of substrate that is utilized for biomass synthesis. The model was suitable for reflecting growth, glucose uptake, and butyric acid time course during fermentation.

6.3.1 Growth models for *C. perfringens*

Besides the aim of describing clostridia cultivations related to production of acids and solvents, models have also been established in some other areas, in which clostridia species have gained importance.

Among these are growth models of *Clostridium perfringens* which is responsible for most cases of foodborne illness. Its spores survive a heat treatment of 100°C for more than an hour, hence it can survive under typical process conditions in ham, roast, and corned beef[65]. Spores that survive in the meat can germinate between 10 and 52°C. In order to predict the effect of temperature on the growth of *C. perfringens*, multiple reports present growth simulations in various forms of meat products. Data of isothermal studies were applied for dynamic growth models. For this purpose, experimental data were fitted by non-linear least-squares regression to a modified Gompertz model. This equation was differentiated against time, yielding a descriptive term for dynamic growth:

$$\frac{dL}{dt} = \mu(L - A)\ln\left(\frac{B - A}{L - A}\right) \tag{6.45}$$

where A represents the initial, and B the final cell concentration. L is the logarithm of the colony forming units (CFU) related to the cell concentration (CFU/c_X) with $L_0 \rightarrow 0$[66]. The simulation is valid for the lag phase, all growth phases, and the stationary phase, respectively. For the prediction of the temperature-dependent growth characteristics, the approach described in Eq. (6.45) was modified[67]. The empirical approach for the description of bacterial growth through a wide temperature range was formulated as follows:

$$\mu = a(T - T_{min})^2\left\{1 - e^{b(T - T_{max})}\right\} \tag{6.46}$$

where a and b represent empiric coefficients. Based on experimental data, these coefficients were fitted with non-linear regression. The presented methodology was applied in a similar approach in another study[68]. Here, the suitability of several empiric correlations for simulating clostridial growth was compared. The already mentioned Gompertz function

$$L(t) = A + (P - A)e^{-e^{-B(t-M)}} \tag{6.47}$$

was applied as well as a logistic function

$$L(t) = A + (P - A)\frac{1}{1 + e^{-B(t-M)}} \tag{6.48}$$

While P and M are empiric and estimated from the data, A was derived to be the average measured population (log CFU) at the time when a minimal population occurred. The application of Eq. (6.48) provided a better approximation when growth was enhanced. The reciprocal of the outgrowth and lag phase duration as well as the exponential growth rate were estimated based on the temperature-dependent growth model[67]. With this approach, the growth of *C. perfringens* at temperatures between 15 and 51°C was simulated. However, the model did underestimate the growth at exponential cooling of ground beef[69].

Several other models are described in the literature, which basically follow a similar methodology in order to describe growth of *C. perfringens* at different temperatures[70,71]. In another approach[72], the pH and the NaCl content in the heat treated meat was also considered as well as the sub-lethal heat damage. Therefore, two independent parameters were introduced into the model equations, which represent the physico-chemical properties of the meat.

An advanced approach was implemented to decrease the deviations between measured and model predicted growth of *C. perfringens* that still remained present in the approaches mentioned before[73]. Growth models were modified, assuming a γ-distribution with different values accounting for unevenly distributed durations of the lag time for each cell. The applied growth models were based on two approaches:

$$n(t) = n(0) + \mu A(t) - \ln\left(1 + \frac{e^{\mu A(t)} - 1}{e^{(m-n(0))}}\right) + \varepsilon(t) \qquad (6.49)^{74}$$

with

$$A(t) = t + \mu^{-1} \ln\left(\frac{e^{-\mu t} + q}{1+q}\right)$$

and a logistic function

$$\mathrm{m(t)} = \mathrm{a} + \mathrm{r}\left(\frac{1}{1+\mathrm{e}^{(\mathrm{j(b-t)})}} - \frac{1}{1+\mathrm{e}^{\mathrm{jb}}}\right) + \varepsilon(\mathrm{t}) \qquad (6.50)^{75}$$

In Eq. (6.49), *n(t)* is the natural logarithm of the number of cells, $\varepsilon(t)$ is the error term, and μ, *m*, and *q* are parameters determining the maximum growth rates. In Eq. (6.50), the parameters *a,b,r*, and φ are estimated parameters. The cell specific distribution of the time in which cells stay in the lag phase was expressed as a γ-distribution:

$$g(t;\alpha,\beta)=\frac{t^{\alpha-1}e^{-t/\beta}}{\beta^{\alpha}\Gamma(\alpha)} \tag{6.51}$$

It was used to simulate lag times for *Listeria innocua*[76]. These models were integrated into secondary temperature-dependent models similar to the previously described methodology:

$$\frac{dm_D(t)}{dt}=h(t)m_0(t)+\mu(t)m_D(t)\times\left(1-\frac{m_0(t)+m_D(t)}{M}\right) \tag{6.52}$$

where $h(t)$ is the hazard function for cells in the lag phase, $m_0(t)$ is the number of cells in the lag phase, and $m_D(t)$ the number of cells in the exponential phase. M represents the maximum population density.

Relative growth could be simulated in a temperature range between –10 and 53°C. In order to improve the prediction quality of real behavior, the history of cells (that is the temperature that cells were exposed to) is assumed to have an impact on the hazard function $h(t)$ of the cell when entering the exponential growth phase and on the specific growth rate $\mu(t)$. For this model extension, an additional parameter Δ was introduced. The functions were assumed to be weighted integrals from $t-\Delta$ to t of the isothermal functions. It was found that a value of $\Delta = 0.5$ was best suited to simulate growth[77].

6.3.2 Fermentative biohydrogen production by clostridia

Another economically interesting application for clostridia species is the field of fermentative biohydrogen production. There is potential for biohydrogen production in *C. acetobutylicum*, *C. butyricum*, *C. tyrobutyricum*, and *C. beijerinckii* cultivations[78]. A kinetic model was applied to describe and evaluate the glucose consumption for these strains. Based on the derived yield coefficients for all species observed, the portion of the consumed glucose which was used for synthesis of a main product was estimated. The hydrogen, acid, and solvent production was predicted and compared to experimental data. The glucose consumption in the utilized model was defined as:

$$r_{Glu} = \frac{q_{Glu}^{max} c_{Glucose} c_X}{K_{Glu} + c_{Glucose}} I_{pH} \tag{6.53}$$

where r_{Glu} represents the glucose consumption, q_{Glu}^{max} the maximum specific glucose consumption, and K_{Glu} the half-saturation constant with respect to glucose. I_{pH} denotes an empiric inhibition term of the form:

$$I_{pH} = e^{-3\left(\frac{pH - pH_{UL}}{pH_{UL} - pH_{LL}}\right)^2} \Bigg|_{pH < pH_{UL}} \tag{6.54}$$

pH_{UL} represents the pH (below pH = 7.0), until which acidogenic bacteria are not inhibited, pH_{LL} denotes the pH at which the reaction is completely inhibited.

Concentrations were estimated by calculating the corresponding part of the glucose uptake which is supporting product synthesis:

$$c_P = (1 - Y_{X/S})\, Y_i \tag{6.55}$$

where Y_i denotes the yield coefficient of product I from glucose. Parameter estimation and model simulations were performed by applying the AQUASIM software package[79]. For all four species, parameter values at the growth and stationary phase in batch cultivations were determined and predicted, respectively.

An empiric approach to simulate hydrogen production by *C. butyricum* following a modified Gompertz equation was applied:

$$H = P \cdot \exp\left(-\exp\left(\frac{R_{max} e}{P}(\lambda - t) + 1\right)\right) \tag{6.56}$$

where H denotes the amount of hydrogen produced in mL, λ the lag time in h, P the biohydrogen production potential in mL, and R_{max} the maximum biohydrogen production rate in mL h^{-1}. This equation was applied to simulate biohydrogen accumulation in cultivations of *C. butyricum* EB6, a strain that was isolated from palm oil mill effluent[80]. Parameters were fitted to P = 4,392 mL, R_{max} = 867 mL h^{-1}, and λ = 5.8 h, yielding a regression of $R^2 > 0.99$.

By means of an experimental design approach, the hydrogen yield was optimized based on the pH range, glucose, and $FeSO_4$ concentration, respectively. Based on the optimization, a model equation that was able to predict the biohydrogen yield depending on process parameters was created. In the study, the model prediction of maximum hydrogen yield at a pH of 5.6, a glucose concentration of 15.7 gL^{-1} and a $FeSO_4$ concentration of 0.39 gL^{-1} was experimentally verified. A similar approach, using response surface methodology, was applied to optimize the hydrogen production of *C. tyrobutyricum* JM1, isolated from a food waste treatment process[81]. The model equation for the hydrogen yield indicated significant interaction of the parameters glucose concentration, pH, and temperature. Optimum conditions were defined to be 103 mM (glucose concentration), a pH of 6.5, and a temperature of 35°C.

Several attempts have been made to describe the biomass growth of cultures of hydrogen-producing bacteria. In another study[82], the model of Eq. (6.52) was applied to describe the hydrogen production of *C. pasteurianum* strain CH4 and *C. butyricum* strain CGS2 on hydrolyzed soluble starch and cassava starch. The maximum hydrogen production H_{max} was estimated to be 6.3 mL h^{-1} by *C. pasteurianum* CH4 and 12.4 mL h^{-1} by *C. butyricum* CGS2, both on hydrolyzed cassava starch. The applicability of empiric growth models as modified Gompertz and Andrew models was reviewed[83]. These approximations were also compared to equations based on Monod kinetics[84,85]. The Andrew and modified Andrew model equations, respectively, for substrate consumption in batch cultivations, were formulated as follows:

$$\frac{dc_S}{dt} = -\frac{1}{Y_{X/S}} \frac{R_{max} c_S}{K_S + c_S - \frac{c_S^2}{K_I}} c_X \tag{6.57}$$

and

$$\frac{dc_S}{dt} = -\frac{1}{Y_{X/S}} \frac{R_{max} c_S}{K_S + c_S + \frac{c_S^2}{K_I}} c_X \tag{6.58}$$

They were able to simulate substrate uptake more precisely than the Monod equation, which did not incorporate any term for substrate inhibition. Some models included inhibitory effects of high salt or hydrogen concentrations in mixed anaerobic fermentations[86]. The inhibitory effect of sodium acetate on the specific sucrose degradation was defined to be:

$$q_S = q_{max} \frac{1}{1 + \left(\frac{c_{Acertate}}{K_{I,Acetate}} \right)^m} \tag{6.59}$$

The inhibitory effects of butyric acid concentrations on the specific growth rate in *C. tyrobutyricum* fed-batch cultivations were described with the following approximation:

$$\mu = \frac{\mu_{max} K_{I,Butyrate}}{K_{I,Butyrate} + c_{Butyrate}} \tag{6.60}$$

Several attempts were made to consider growth and hydrogen yield to simulate hydrogen production in a Luedeking–Piret type model equation[83]. Growth of *C. butyricum* strain CGS5 on xylose applying the Monod equation (R^2 = 0.881) was described[87]. With the same approximation, growth of *C. pasteurianum* strain CH4 on sucrose was modeled (R^2 = 0.970). The hydrogen yield with respect to biomass growth ($Y_{H2/X}$) was estimated using a modified Luedeking–Piret model:

$$\frac{1}{c_X} \frac{dc_{H2}}{dt} = Y_{H2/X} \left(\frac{1}{c_X} \frac{dc_X}{dt} \right) \tag{6.61}$$

The hydrogen yield for the xylose fermentation could be estimated to be 0.041 mol g^{-1} (*VSS* = volatile suspended solids), and for the sucrose fermentation to 0.039 mol g^{-1} (*VSS*) ($R^2 > 0.910$ for both simulations).

6.4 Outlook

The models presented in this report have been developed for different purposes to simulate cultivation processes with clostridia. Most of them were created independently from each other, every model covering only a small section of the complex field regarding clostridia. Improved data and model management could create more widely applicable model approaches by collecting the numerous modeling attempts, evaluating their validity with identical data sets, and combining them in an enhanced model.

New insights into the regulation of the cell or knowledge of signal molecules of major importance should be used to reduce empirical model parameters as much as possible. Hence, the increasing observations on the different regulatory levels of the cell should be included in process models as far as they do not become impractical due to high complexity and specificity. The combination of process model approaches and systems biology approaches does not only result in a better prediction capability of process models. It will also result in a better understanding of how process design and regulation should be performed to supply the demands of the cell for optimal production. First attempts are made with the development of a genome-scaled model[41,88] that is integrated in solving flux balance equations and applying this methodology to predict key process parameters[42].

A case-dependent (that is dependent on the state of the process) selection of modeling modules would support a broader range of possible applications. For example, it is unlikely that a small number of factors or even a single molecule is responsible for the switch from the acidogenic growth phase to the solventogenic phase in the ABE fermentation. At this point, a single comprehensive model is not able to cover all possible situations that are relevant for multiple changes of the process and physiologic characteristics usually observed. Recognition of the key mechanisms and selection of the related modeling modules can reduce the complexity of models. This should be sufficient for the application in process development and control.

Not only the continuous improvement of prediction quality is of importance, but also the combination of models with process analytical tools. In case of the ABE fermentation, conventional and new methods have been examined for the utilization as *on line* or *at line* analytical tools[89–91]. The integration of these methods into a model-based process control strategy will provide better process stability, increasing the economic viability, and hence secure competitiveness of clostridia fermentations.

Nomenclature

Parameter	Unit	Description
a_c	m^{-1}	cell volume specific surface area
a_{ij}	-	stoichiometric coefficient of species i in pathway j
A	-	stoichiometric matrix
c_i or c_{int}	mol/cell volume	intracellular concentration
c_{AA}	mol_{AA}/reactor volume	total extracellular acetic acid concentration
c_{BA}	mol_{BA}/reactor volume	total extracellular butyric acid concentration
c_{AH}	mol_{AH}/reactor volume	extracellular concentration of undissociated acid
c_{A^-}	mol_{A-}/reactor volume	extracellular concentration of dissociated acid
c_{AHi}	mol_{AH}/cell volume	intracellular concentration of undissociated acid
c_{A^-i}	mol_{A-}/cell volume	intracellular concentration of dissociated acid
$c_{a,d}$	$mol_{a,d}$/reactor volume	extracellular concentration of dissociated acid a
$c_{a,un}$	$mol_{a,un}$/cell volume	extracellular concentration of undissociated acid a
$c_{a,i,d}$	$mol_{a,d}$/cell volume	intracellular concentration of dissociated acid a
$c_{a,i,un}$	$mol_{a,un}$/cell volume	intracellular concentration of undissociated acid a
c_{inh}	mol/reactor volume	inhibitor concentration
c_P	mol/reactor volume	product concentration
c_s	gL^{-1}	substrate concentration
c_T	mol/reactor volume	total concentration of active sites
c_x	gL^{-1}	dry biomass concentration
c_I	mol/reactor volume	extracellular concentration of species I
K	h^{-1}	kinetic constant
K_i	H^{-1}	inhibition constant
k_{un}		transport coefficient for undissociated acids
k_d		transport coefficient for dissociated acids
l	-	lower limit vector at objective function for flux estimation

(*Continued*)

μ	h^{-1}	specific growth rate
μ_i	h^{-1}	specific growth rate at inhibition
μ_{max}	h^{-1}	maximum specific growth rate
pHi	-	intracellular pH
q_{BUOH}	h^{-1}	Specific butanol release rate
q_S	h^{-1}	Specific substrate uptake rate
ρ_x	kg m^{-3}	density of the cell
r_j	mol/(reactor volume·h)	reaction volume specific rate of reaction j
r_A	mol·L^{-1}·h^{-1}	cell volume specific extracellular formation of total acids
r_{AH}	mol·L^{-1}·h^{-1}	cell volume specific extracellular formation of undissociated acids
r_{ACUP}	mol·L^{-1}·h^{-1}	cell volume specific acetic acid uptake flux
r_{BYUP}	mol·L^{-1}·h^{-1}	cell volume specific butyric acid uptake flux
$r_{BA,uptake}$	mol·L^{-1}·h^{-1}	rate for conversion of butyric acid to butanol
r_i	mol·L^{-1}·h^{-1}	cell volume specific intracellular rate
R	mol·h^{-1}	Total conversion rate
t	s	Time
t_N	s	Time
u	-	upper limit vector
v_n	-	vector of stoichiometric coefficients
V_B	L	cell volume
W	-	weighting matrix
y	-	culture physiology state marker
$Y_{X/S}$	-	biomass yield coefficient based on sugar substrate consumption
x	-	vector of metabolite concentrations
x_{int}	mol/cell volume	concentration of intermediates

References

1. Killeffer D. H., *Ind. Eng. Chem.* 19 (1927).
2. Jones D. T. and D. R. Woods., *Microbiol. Rev.* 50 (1986).
3. Robinson G. C., *J. Biol. Chem.* 53 (1922).
4. Weinstein L. and Rettger L .F., *J. Bacteriol.* 32 (1932).
5. Weyer E. R. and Rettger L. F., *J. Bacteriol.* 14 (1927).
6. Bilford H. R. *et al., Ind. Eng. Chem.* 34 (1942).
7. Borzani W., *Engenharia Quimica* 1 (1953).
8. Gaden E. L., *Chem. Ind.* 7 (1955).
9. Borzani W., *Agr. Food. Chem.* 5 (1957).
10. Borzani W. *et al., Appl. Microbiol.* 8 (1960).

11. Ueda K. *et al., J. Agr. Chem. Soc.* 28 (1954).
12. Nagatani M. *et al., J. Ferment. Technol.* 46 (1968)
13. Maxon W. D., *Appl. Environ. Microbiol.* 3 (1955).
14. Kono T. and Asai T., *J. Ferment. Technol.* 47 (1969).
15. Erickson L. E. *et al., Biotech. Bioeng.* 20 (1978).
16. Wood W. A., in *The Bacteria*, Gunsalus I. C. and Stanler R. Y., Eds., Chapter 2 (Academic Press, New York, 1961), pp. 59–149.
17. Papoutsakis E. T., *Biotech. Bioeng.* 26 (1983).
18. Yang X. and Tsao G. T., *Biotechnol. Prog.* 10 (1994).
19. Votruba J. *et al., Biotech. Bioeng.* 28 (1986).
20. Yerushalmi L. *et al., Biotech. Bioeng.* 28 (1986).
21. Shinto H. *et al., J. Biotechnol.* 131 (2007).
22. Desai R. P. *et al., J. Biotechnol.* 71 (1999).
23. Wiesenborn D. P. *et al., Appl. Environ. Microbiol.* 55 (1989).
24. Desai, R. P. and E. T. Papoutsakis. *Appl. Environ. Microbiol.* 65 (1999).
25. Harris, L. M. *et al., J. Bacteriol.* 184 (2002).
26. Sillers R. *et al., Biotech. Bioeng.* 102 (2009).
27. Sillers R. *et al., Metab. Eng.* 10 (2008).
28. Tummala S. B., *et al., J. Bacteriol.* 185 (2003).
29. Junne S., Ph.D. Thesis, TU Berlin, Germany (2010).
30. Jones W. *et al., Gen. Biol.* 9 (2008).
31. Huang L. *et al., Appl. Environ. Microbiol.* 52 (1986).
32. Huang L. *et al., Appl. Environ. Microbiol.* 50 (1985).
33. Huesemann M. and Papoutsakis E. T., *Abstracts of Papers of the American Chemical Society* 194 (1987).
34. Zhao Y. *et al., Appl. Biochem. Biotechnol.* 71 (2005).
35. Marquet M. *et al., Biochimie* 60 (1978)
36. Gheshlaghi R. *et al., Biotechnol. Adv.* 27 (2009)
37. Uyeda K. *et al., J. Bacteriol. Chem.* 246 (1971)
38. Fromm H.J., *Initial rate enzyme kinetics* (Springer, Berlin 1975)
39. Waterson R.M. *et al., J. Biol. Chem.* 247 (1972)
40. Schaupp A. *et al., Arch. Microbiol.* 100 (1974)
41. Senger R. S. and Papoutsakis E .T., *Biotech. Bioeng.* 101 (2008).
42. Senger R. S. and Papoutsakis E. T., *Biotechnol. Bioeng.* 101 (2008).
43. Ni Y. and Sun Z., *Appl. Microbiol. Biotechnol.* 83 (2009).
44. Vanandel J. G. *et al., Appl. Microbiol. Biotechnol.* 23 (1985).
45. Heyndrickx M. *et al., Biotechnol. Lett.* 12 (1990).
46. Heyndrickx M. *et al., Enz. Microbial. Technol.* 13 (1991).
47. Abbad-Andaloussi S. *et al., Microbiol.-UK* 142 (1996).
48. Guedon E. *et al., J. Bacteriol.* 181 (1999).
49. Desvaux M. and Petitdemange H., *Microb. Ecol.* 43 (2002).
50. Gapes J. R. *et al., Appl. Environ. Microbiol.* 62 (1996).

51. Mutschlechner O. *et al., J. Mol. Microbiol. Biotechnol.* 2 (2000).
52. Tashiro Y. *et al., J. Biotechnol.*120 (2005).
53. Srivastava A. K. and Volesky B., *Biotech. Bioeng.* 38 (1991).
54. Mulchandani A. and Volesky B., *Can. J. Chem. Eng.* 64 (1986).
55. Luedeking R. and Piret E. L., *J. Biochem. Microbiol. Technol. Eng.* 1 (1959).
56. Soni B. K. *et al., Biotech. Bioeng. Symp.* 17 (1986).
57. Jarzebski A. B. *et al., Bioprocess. Eng.* 7 (1992). **7**: p. 357–361.
58. Herrero A. *et al., Appl. Microbiol. Biotechnol.* 22 (1985).
59. Janssen H. *et al., Appl. Microbiol. Biotechnol.* 87 (2010).
60. Haus S *et al., BMC Syst. Biol.* 5 (2011).
61. Hill P.W. *et al., Biotech. Bioeng.* 43 (1993).
62. Biebl H., *Appl. Microbiol. Biotechnol.* 35 (1991).
63. Zeng A.P. *et al., Biotech. Bioeng.* 44 (1994).
64. Kong Q. *et al., Lett. Appl. Microbiol.* 43 (2006).
65. Barnes E. M. *et al., J. Appl. Microbiol.* 26 (1963).
66. Huang L., *Int. J. Food. Microbiol.* 87 (2003).
67. Ratkowsky D. A. *et al., J. Bacteriol.* 154 (1983).
68. Juneja V. K. *et al., Food Microbiol.* 16 (1999).
69. Smith S. and Schaffner D. W., *Appl. Environ. Microbiol.* 70 (2004).
70. Amezquita A. *et al., Int. J. Food Microbiol.* 101 (2005).
71. Smith S. and Schaffner D. W., *J. Food Prot.* 68 (2005).
72. Le Marc Y. *et al., Int. J. Food. Microbiol.* 128 (2008).
73. Juneja V. K. *et al., Food Microbiol.* 25 (2008).
74. Baranyi J. *et al., Food Microbiol.* 10 (1993).
75. Corradini M. G. *et al., Int. J. Food Microbiol.* 106 (2006).
76. Metris A. *et al., J. Microbiol. Methods* 55 (2003).
77. Juneja V. K. *et al., In. Food Sci. Em. Technol.* 10 (2009).
78. Lin P. Y. *et al., Int. J. Hydr. Energy* 32 (2007).
79. Reichert P. *et al., AQUASIM 2.0.* (Swiss Federal Institute for Environmental Science and Technology, Duebendorf, Switzerland, 1998).
80. Chong M. L. *et al., Int. J. Hydr. Energy* 34 (2009).
81. Jo J. H. *et al., Int. J. Hydr. Energy* 33 (2008).
82. Chen S. D. *et al., On1.* (2007).
83. Wang J. and W. Wan. *Int. J. Hydr. Energy* 34 (2009).
84. Kumar N. *et al., Int. J. Hydr. Energy* 25 (2000).
85. Nath K. *et al., Int. J. Hydr. Energy* 34 (2008).
86. Wang Y. *et al., Int. J. Hydr. Energy* 33 (2008).
87. Lo Y. C. *et al., Water Res.* 42 (2008).
88. Lee J. *et al., Appl. Microbiol. Biotechnol.* 80 (2008).
89. Chauvatcharin S. *et al., J. Ferment. Bioeng.* 79 (1995).
90. Junne S. G. *et al., Biotech. Bioeng.* 99 (2008).
91. Kansiz M. *et al., Anal. Chim. Acta.* 438 (2001).

CHAPTER 7

MODELLING *AGR*-DEPENDENT QUORUM SENSING IN GRAM-POSITIVE BACTERIA

S. JABBARI

School of Mathematics
University of Birmingham
Edgbaston Birmingham, B15 2TT, UK

J. R. KING

Division of Theoretical Mechanics,
School of Mathematical Sciences University of Nottingham
Nottingham, NG7 2RD, UK

N. P. MINTON, K. WINZER

School of Molecular Medical Sciences,
Centre for Biomolecular Sciences University of Nottingham
Nottingham, NG7 2RD, UK

Many bacteria are known to regulate genes in concert with cell density, a phenomenon termed 'quorum sensing'. Stochastic and deterministic modelling approaches have revealed previously unknown properties of quorum sensing systems and generated novel, experimentally testable hypotheses. This chapter focuses on *agr*-type quorum sensing systems, which rely on the production of an autoinducing cyclic peptide signal that is sensed via a classical two-component system. *agr*-dependent quorum sensing is currently best understood in staphylococcal species, but homologous systems are present in a wide range of Gram-positive bacteria, particularly members of the class *Clostridia*. Deterministic modelling approaches for the staphylococcal *agr* systems are presented. Using *Clostridium acetobutylicum* as an example, it is investigated how existing models can be adapted to investigate *agr* systems present in other species.

7.1 Quorum sensing: gene regulation in response to cell population density

Many bacteria are known to produce small, diffusible signal molecules. It is often assumed that these molecules are employed to coordinate gene expression with cell population density, a phenomenon known as 'quorum sensing'[1,2]. The underlying idea is that signal molecules are constitutively produced by growing cell populations and can thus act as a measure for their density. Once a critical signal molecule concentration has been achieved, a response is triggered that leads to a wide range of transcriptional and phenotypic changes. Bacteria are thus able to communicate with each other and to coordinate population-wide responses.

Regarding the chemical nature of the signal molecules employed, Gram-negative and Gram-positive bacteria appear to differ in their preferences. In Gram-negative bacteria, signal molecules often contain acyl side chains, such as those present in the extensively studied *N*-acyl homoserine lactones (AHLs), the *Pseudomonas* quinolone signal (PQS) or hydroxyl-palmitic acid methyl ester (PAME)[1]. The known Gram-positive systems, however, are mostly based on secreted peptides which can be either linear or cyclic, or contain other extensive posttranslational modifications[3-5].

Quorum sensing mechanisms are involved in the regulation and fine-tuning of a large and diverse range of phenotypic traits. For Gram-positive bacteria, specific examples include genetic competence (*Bacillus subtilis*, *Streptococcus* spp.)[3,4,6-8], conjugation (*Enterococcus faecalis*)[9,10], sporulation (*B. subtilis*)[4], production of bacteriocins, lantibiotics and other antimicrobials (*Lactococcus* spp., *Lactobacillus* spp., *Carnobacterium piscicola*, *B. subtilis*, *Enterococcus faecium*, *Streptococcus thermophilus*)[3,11,12], regulation of virulence factors including exoenzymes and toxins (*Clostridium perfringens, E. faecalis*, *Listeria monocytogenes*, *Staphylococcus* spp., *Bacillus cereus* group)[3,5,13-17], as well as biofilm formation (*Staphylococcus* spp., *Streptococcus mutans, L. monocytogenes*)[5,16,18].

7.2 *agr*-type quorum sensing systems

This chapter focuses on *agr*-homologous systems, so far found exclusively in the phylum Firmicutes. Although best understood in staphyloccocal species, in particular *Staphylococcus aureus*[5], these systems appear to be widely distributed among members of the class clostridia (see below), including *Clostridium acetobutylicum*.

The staphylococcal *agr* system consists of four proteins encoded by a single operon, *agrBDCA*, as well as a divergently transcribed regulatory RNA, RNAIII (Fig. 7.1(a))[5]. A so-called autoinducing peptide (AIP) acts as a signal molecule

AgrA. Binding of the AIP to external loops of AgrC results in phosphorylation of AgrA. In this activated form, AgrA (along with additional regulatory proteins such as SarA) is able to bind to the P2 promoter upstream of *agrBDCA* (Fig. 7.1(a)) thus stimulating transcription of the entire *agr* operon and, thereby, creating a typical autoactivation circuit. As a consequence, the intracellular concentration of activated AgrA rapidly increases, leading to changes in quorum sensing-dependent gene expression. Until recently, the regulation of target genes in response to phosphorylated AgrA levels was thought to be mediated exclusively by RNAIII. Indeed, up- and downregulation of many toxin and surface protein genes is dependent on this RNA, the concentration of which drastically increases due to direct activation of the P3 promoter by AgrA once the population is 'quorate'. However, many metabolic genes as well as the *psm* cytolysins are now known to be regulated by AgrA independently of RNAIII[19]. *agr*-homologous systems have also been described and experimentally studied in several other bacteria, including *E. faecalis* (*frs* system), *Lactobacillus plantarum* (*lamBDCA*), *L. monocytogenes* and *C. perfringens*[5,13,15-17,20-24]. With exception of the latter, all these systems contain a complete *agr* locus, i.e. a cluster of *agrBDCA*-homologous genes. In *C. perfringens*, the *agrBD* genes are located elsewhere and transcribed independently of the genes required to detect and transduce the putative (but as yet unidentified) AIP signal, the *virRS* two-component system[17,24]. No RNAIII homologues have as yet been identified outside of the genus *Staphylococcus*. However, in *C. perfringens* phosphorylated VirR

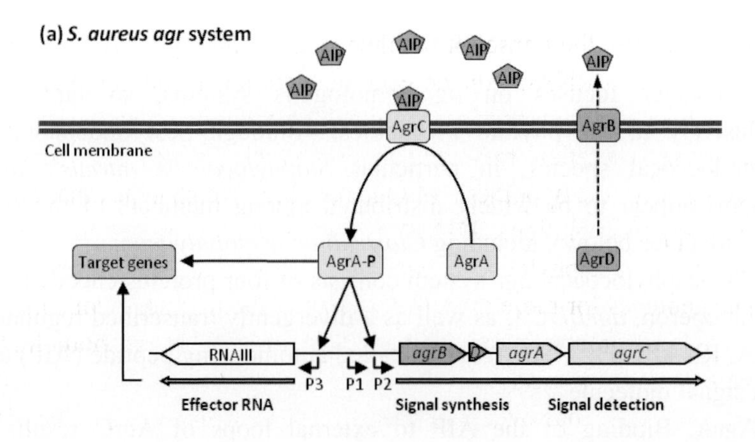

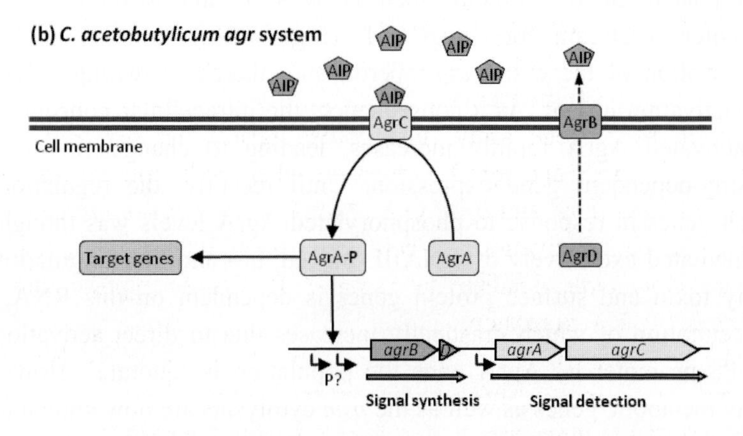

Fig. 7.1: *agr* quorum sensing circuits. The *agr* locus comprises a single operon in *S. aureus* (a) and two transcriptional units, *agrBD* and *agrCA*, in *C. acetobutylicum* (b). AgrB is involved in the processing of AgrD, resulting in mature AIPs. Binding of AIPs to membrane-located histidine kinase AgrC triggers a conformational change that allows phosphorylation of the response regulator AgrA. Straight solid arrows: regulation of gene expression; curved arrows: AgrA phosphorylation; dashed arrows: production of mature AIP from AgrD; arrows labelled P, P1, P2, P3: promoters; grey-filled arrows: *agr* and RNAIII gene transcripts. Interactions and promoters in (b) are hypothetical.

directly activates the transcription of several regulatory RNAs[25] and thus it is possible that RNA-mediated gene regulation is a common feature of *agr*-dependent quorum sensing circuits. However, for some *agr* systems target gene regulation may rely solely on phosphorylated AgrA, in agreement with the hypothesis of Queck *et al.*[19] which states that during the course of evolution, RNAIII-mediated control was added to an already existing staphylococcal *agrBDCA* quorum sensing system.

The genomes for more than 220 species of the phylum Firmicutes are now available. Interestingly, their analysis reveals that *agr*-homologous systems exist in many of them and particularly in the class *Clostridia*[17,24,26,27] (our own unpublished data), sometimes in the form of complete *agrBDCA* clusters.

C. acetobutylicum is one of the few currently sequenced members of the genus that appear to possess a complete *agrBDCA* cluster (Fig. 7.1(b)). In contrast to the staphylococcal system, however, *agrBD* is predicted to be transcribed independently of *agrCA*[28], and this is supported by a gene array time course series that revealed strong increases in *agrB* and *agrD* expression when cells entered stationary phase, whereas *agrA* and *agrC* expression remained relatively constant[a]. Thus, an autoactivation circuit may also be operational in *C. acetobutylicum*, leading to rapid AIP signal amplification via increased *agrBD* expression.

The genes and phenotypes regulated by the *C. acetobutylicum agr* system are currently under investigation in the authors' laboratory. Intriguingly, the number of heat-resistant endospores that are formed by *agrB, agrD, agrC* and *agrA* mutants is reduced when compared to the wild-type suggesting that, similar to *B. subtilis*, cell density-dependent signalling may be involved in the regulation of endospore formation. However, important differences between *Bacillus* and *Clostridium* are thought to exist at the level of sporulation initiation. In *B. subtilis*, initiation of endospore formation is absolutely dependent on nutrient limitation, whereas *C. acetobutylicum* seems to be able to sense the potentially life-threatening conditions generated through its fermentation

[a] Our analysis of supplemental gene array data presented by Alsaker and Papoutsakis[29].

metabolism in an as yet unknown way and is thus able to sporulate even in the presence of excess nutrients[30,31]. Cell density-mediated control of sporulation in *B. subtilis* is mediated by several small, linear peptides, PhrA, CSF and PhrC[4], but high cell density alone is not sufficient to initiate sporulation; rather, quorum sensing is thought to modulate the numbers of *B. subtilis* cells that will sporulate under conditions of starvation. The exact role of cell density signals in *C. acetobutylicum* sporulation remains to be established, but the fact that in liquid culture sporulation of *agr* mutants is not abolished suggests that, as in *Bacillus*, the system plays a modulatory role in conjunction with additional regulatory operons, forming part of a much larger gene regulation network governing sporulation.

A purely experimental approach to understanding such networks becomes increasingly difficult as additional components are revealed. Mathematical models provide a useful tool with which to examine these networks, either isolating specific interactions of particular significance, or investigating the overall network and its influence upon the population-wide response. The remaining sections of this chapter will review various modelling approaches that have been developed for bacterial quorum sensing systems and discuss how they could be adopted for *agr* systems present in clostridia; in particular we consider why regulation of the *agr* system in *C. acetobutylicum* deviates from the standard form observed in *S. aureus*.

7.3 Mathematical models of quorum sensing

Quorum sensing was first modelled mathematically in Gram-negative bacteria[32-35] and the majority of existing models still focus on the *lux* system and its homologues which are prevalent in such bacteria (for instance *lux* in *Vibrio fischeri* or *las* and *rhl* in *Pseudomonas aeruginosa*). Models range from subcellular approaches, wherein the relevant gene regulation network is captured explicitly (for example James *et al.* [32], Dockery and Keener [33] and Fagerlind *et al.*[36]), to the cellular and population levels where the focus is on the proportion of cells in each state of quorum sensing-induced regulation and on the concentration of the signal molecule (for example Ward *et al.*[34] and

Nilsson *et al.*[35]). While the gene-regulation models mentioned above suggest that the quorum sensing systems enable bistable behaviour to arise, Ward *et al.*[34] demonstrate that quorum sensing could instead involve the proportion of upregulated cells increasing steadily as the quorum sensing signal accumulates. When bistability is present hysteresis occurs, an abrupt switch to the quorate state occurring at a higher population density in a growing population than that at which the switch to the inquorate state takes place in one that is becoming more dilute; in the alternative scenario, no hysteresis occurs and the transition between up- and downregulated states is gradual, albeit rapid around some 'critical' level of quorum sensing molecules. Thus, the properties of the system can be dependent upon both the parameter choice and the particular scenario which is being modelled; indeed, this is consistent with later results of Anguige *et al.*[37], Haseltine & Arnold[38] and Kuttler & Hense[39], where it is illustrated how the number of feasible steady states can be dependent upon the parameter values and the number of feedback loops within the quorum sensing circuitry.

The focus of Anguige *et al.*[37] is on repression of the signalling system and this is a common theme in quorum sensing modelling: Anguige *et al.*[40,41] and Fagerlind *et al.*[42] all also consider inhibiting the quorum sensing circuitry for therapeutic purposes since quorum sensing is often linked with the virulence capabilities of bacteria, while Ward *et al.*[43] examine the concept that bacteria use self-repression as a means to optimise efficiency. Other scenarios in which quorum sensing is modelled include the use of the signalling system in a burn wound[44] or in a biofilm (for example Nilsson *et al.*[35], Chopp *et al.*[45] and Ward *et al.*[46]); all of these studies focus on the bacterium *P. aeruginosa*. Additionally, models can be used to test specific hypotheses, such as the influence of protein degradation rates[47] or cell size[48] on quorum sensing-system behaviour, or issues of interpretation, such as the debate regarding whether quorum sensing is a method to detect population number, environmental conditions or a combination of the two[49]; this latter study proposes the term 'efficiency sensing' as a more suitable name for the cell–cell signalling mechanism.

Though some of the above developments may also be applicable to Gram-positive bacteria, quorum sensing models specific to

Gram-positives are, thus far, significantly fewer in number. Koerber *et al.*[50] model the use of quorum sensing by *S. aureus* in endosome escape: a *S. aureus* cell has the ability to become internalised within the endosome of a host cell, it believed that the virulence factors required for endosome escape are under the control of the *agr* operon. Since there is typically only one bacterium contained within an endosome, the authors adopt a stochastic approach and use asymptotic methods and Monte Carlo simulations to approximate the distribution of endosome escape times (examples of stochastic models of quorum sensing in Gram-negative bacteria include Goryachev *et al.*[51,52], Wang *et al.*[53], Hong *et al.*[54] and Müller *et al.*[55]; the focus of the last of these is on how a cell can overcome stochastic effects). Gustafsson *et al.*[5], Jabbari[57] and Jabbari *et al.*[58,59] model (deterministically) the *agr* operon in *S. aureus* at the subcellular level, by developing gene-regulatory network models (this will also be the focus of the subsequent sections of this chapter). Gustafsson *et al.*[56] (2004) consider the influence of the regulatory protein SarA upon activation of the *agr* operon, performing steady-state analyses to demonstrate that this protein can change the sensitivity of a cell to the AIP concentration (a lower bacterial density is required to induce upregulation if SarA-induced P2 activity is higher). Jabbari *et al.*[58], on the other hand, concentrate on the time-dependent dynamics of the *agr* operon, conducting an asymptotic analysis to identify the roles of the various mechanisms involved in the full loop, while Jabbari *et al.*[59] examine the possibility that the phosphorylation cascade between AgrA and AgrC does not follow the classical form outlined in Section 2. Other models of quorum sensing in Gram-positive bacteria include studies of the linear peptide-based quorum sensing system in *B. subtilis* and its influence upon the initiation of sporulation[60,61], again focusing on gene regulation.

The remainder of this chapter returns to the *agr* circuit and discusses the general methodology behind the existing models of gene regulation in the *agr* operon and the motivation for their development. It is hoped that such approaches will also aid the understanding of quorum sensing processes in *C. acetobutylicum*.

7.4 Model formulation

The number and diversity of both cellular and environmental factors impacting upon cell behaviour renders a comprehensive representation of a bacterium (or indeed any organism) outside of the scope of a traditional mathematical model. Each additional feature included in a model brings with it further parameters that must be measured or estimated and, in general, also increases the complexity of the analysis. Consequently, a number of assumptions are essential for any mathematical model; in Table 7.1 we list some of the most common of these for models of gene regulation networks.

In this chapter, we focus on the main principles behind deterministic ordinary differential equation (ODE) modelling. We therefore consider a population of cells in a well-mixed environment where we assume that cell number and molecular copy numbers are sufficiently high that we can neglect stochastic effects. For simplicity we also take the population size to be constant and, given that ribosomes are required for translation of all proteins and not just those of the *agr* system, we assume that a plentiful supply of the former is present. Gustafsson *et al.*[56] do not track mRNA levels, assuming that any gene regulation impacts directly upon protein concentration. Conversely, in Jabbari *et al.*[58,59] mRNA is included in the model to make the gene regulation more explicit.

Table 7.1: A list of common assumptions used in mathematical modelling of gene regulation networks and some of their more straightforward implications.

Assumption	Why?	What is required if not?
Well-mixed environment	Can use ordinary differential equations	Partial differential equations
High copy number of all molecules	Can use deterministic modelling	Stochastic equations
Constant population size	Simplifies the equations	Additional growth term
Plentiful supply of ribosomes	Reduces number of equations	Extra equations to track ribosome number
mRNA levels are quasi steady	Reduces number of equations	Extra equations to track mRNA levels

Certain assumptions must also be made which are specific to the staphylococcal *agr* system. We list here the main ones that we adopt:

- The rates of translation of each of the Agr proteins are the same and proportional to the concentration of *agr* mRNA
- Housekeeping phosphatases act on phosphorylated AgrA
- After modification of AgrD into AIP, we can neglect what remains of the AgrD protein from the model
- Receptor-bound AIP can separate spontaneously
- AgrC phosphorylation occurs as soon as it binds to AIP
- Following phosphate transfer to AgrA, AgrC can autophosphorylate again

In Fig. 7.2(a) we represent schematically each component of the *agr* loop which will be included in the model (additionally, those components which are neglected are depicted by dotted boxes).

In many cases, gene-regulatory-network models are built using conventional mass-action kinetics: the rate of a reaction is taken to be proportional to the product of the concentrations involved in the reaction. Thus, for instance, AgrA is produced via translation of mRNA. If A is the concentration of AgrA and M that of mRNA, we can represent their rates of change by the following ODEs:

$$\frac{\mathrm{dM}}{\mathrm{dt}} = 0, \qquad \frac{\mathrm{dA}}{\mathrm{dT}} = \kappa\mathrm{M} \tag{7.1}$$

mRNA is assumed not to be lost during the process of translation so its concentration remains constant (we consider translation in isolation here; in reality other reactions, such as degradation of mRNA, would affect its concentration) and AgrA is translated from mRNA at rate κ. The same approach applies to more complex reactions such as transmembrane AgrC + AIP → AIP-bound AgrC.

This can be represented by

$$\frac{dR}{dt} = -\beta Ra, \qquad \frac{\mathrm{dR}}{\mathrm{dt}} = -\beta\mathrm{Ra}, \qquad \frac{\mathrm{dR}^*}{\mathrm{dt}} = \beta\mathrm{Ra} \tag{7.2}$$

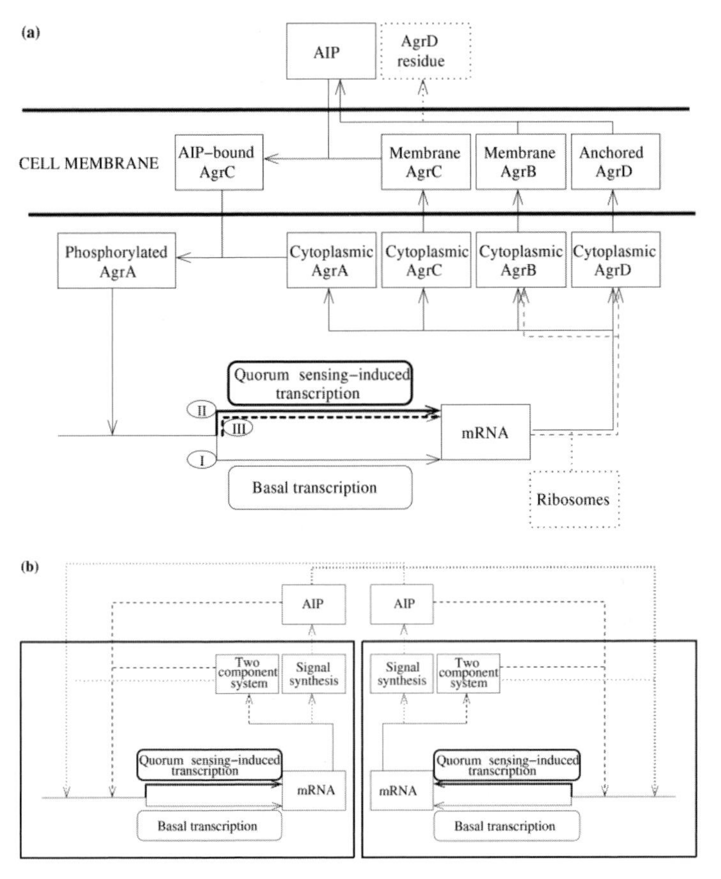

Fig. 7.2: (a) A schematic representation of the *agr* circuitry. The quorum sensing system is governed by two levels of transcription: a relatively low basal level (arrow I) and a higher quorum sensing-induced level (arrow II). In *S. aureus*, each of the four *agr* genes are subject to quorum sensing induction, while in *C. acetobutylicum* this is only the case for *agrB* and *agrD*; thus quorum sensing-related transcription for *C. acetobutylicum* is represented by the dashed arrows (arrow III). Components which are neglected from the model are contained within dotted boxes. (b) An overview of cell–cell communication via the *agr* operon. In *S. aureus* the circuitry can be divided into two subloops, each including two proteins: one induces production of the proteins required for signal synthesis (AgrB and AgrD, dotted line), enabling cell–cell communication, and the other those required for signal detection (AgrA and AgrC, dashed line) leading to activation of the cellular response (for *S. aureus* this is the upregulation of virulence-factor production while for *C. acetobutylicum* this is likely to be the induction of sporulation, amongst other processes). In *C. acetobutylicum* only the signal synthesis subloops operate.

where β is the rate of binding between AgrC and AIP (see Tables 7.2 and 7.3 for definitions of the parameters and variables).

Table 7.2: Definitions of the parameters.

Parameter	Rate constant for	Units
m	Basal production of mRNA	Molecules cells^{-1} s^{-1}
v	mRNA transcription	Molecules cells^{-1} s^{-1}
κ	Protein translation	s^{-1}
α_T, α_R	AgrB and AgrC taken up into cell membrane	s^{-1}
α_S	AgrD anchors to cell membrane	s^{-1}
λ_X	Natural degradation of variable $X(t)$	s^{-1}
r	Dilution through cell division	s^{-1}
δ_X	Degradation and dilution ($\delta_X=\lambda_X+r$)	s^{-1}
k	AIP production from AgrD, mediated by AgrB	Molecules^{-1} cm^3 s^{-1}
β	Binding of AIP to AgrC	Molecules^{-1} cm^3 s^{-1}
γ	Separation of AIP from AgrC	s^{-1}
ϕ	Activation of AgrA by AIP-bound AgrC	Molecules^{-1} cm^3 s^{-1}
μ	Dephosphorylation of AgrA by phosphatases	s^{-1}
b	Binding of the promoter site	Molecules^{-1} cells s^{-1}
u	Unbinding of the promoter site	s^{-1}
N	Total number of bacteria per unit volume	Cells cm^{-3}

Table 7.3: Definitions of the variables.

Variable	Concentration of	Units
M	mRNA	Molecules cm^{-3}
A, B, C, D	Cytoplasmic AgrA, AgrB, AgrC, AgrD	Molecules cm^{-3}
T, R	Transmembrane AgrB, AgrC	Molecules cm^{-3}
S	Anchored AgrD	Molecules cm^{-3}
a	Free AIP	Molecules cm^{-3}
R^*	AIP-bound receptor	Molecules cm^{-3}
A_p	Phosphorylated AgrA	Molecules cm^{-3}
P	Proportion of cells that is upregulated	Dimensionless

Each variable in the model is therefore associated with an ODE that contains terms representing each of the reactions in which that variable is involved. For example, (non-phosphorylated) AgrA can arise as a result of translation from mRNA or housekeeping dephosphorylation of phosphorylated AgrA. Conversely, it can be lost through phosphorylation by AIP-bound AgrC, or through degradation. The right-hand side of the

ODE associated with non-phosphorylated AgrA must therefore contain four terms, each representing one of these reactions: taken together these determine the net rate of change of the level of non-phosphorylated AgrA, represented mathematically by its time derivative constituting the left-hand side of the ODE.

In order to incorporate gene regulation into the model we assume that *agr* mRNA is transcribed at two rates: at a basal rate, m, and at a higher rate, v, induced by phosphorylated AgrA; the latter rate is taken to increase linearly with the average level of upregulation in the population, P, and for *C. acetobutylicum* applies only to *agrBD* (in the case of *C. acetobutylicum* transcription of *agrCA* is assumed to occur at the basal rate only).

Mathematical models should be geared towards their aims, e.g. it is possible to select only the elements of the biological circuit that are of most interest and render the analysis simpler. For example, Gustafsson *et al.*[56] concentrate on the two-component system and add an additional component, the regulatory protein SarA, in order to examine the influence of SarA upon *agr* activity. For completeness, in Fig. 7.3 we provide a set of equations that represents every element of the staphylococcal *agr* circuitry; this enables examination of the full feedback loop that is believed to be central to the nature of quorum sensing in *S. aureus*.

One of the major drawbacks of gene regulation modelling is the large number of parameters which result from incorporating all the protein interactions combined, often with an absence of complementary experimental data from which estimates for such parameter values can be obtained. When parameter values are unknown, one possibility is to exploit asymptotic techniques where, rather than estimating absolute parameter values, we estimate relative sizes of parameters; this suffices for qualitative investigations and allows the dominant mechanisms to be identified. Comparison of parameters in this sense requires that they be of the same dimension and therefore nondimensionalisation of the system of equations is an important step; this involves scaling variables with quantities having the same dimensions, i.e. concentrations are scaled with representative levels and so on (see Jabbari *et al.*[58] for more details).

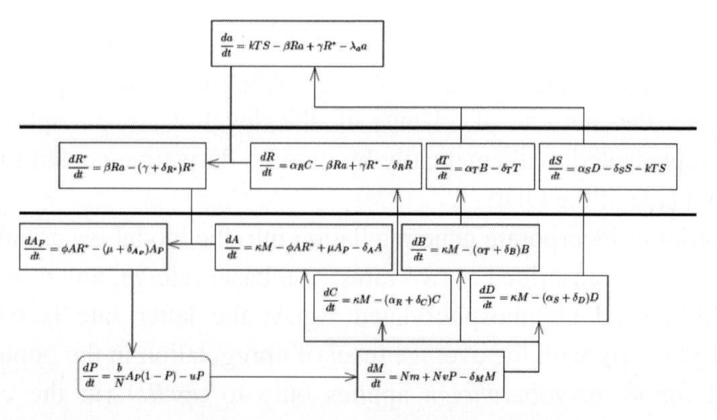

Fig. 7.3: A schematic representation of the mathematical model of the staphylococcal *agr* circuit (compare Fig. 7.2 (a)). See Tables 7.2 and 7.3 for definitions of parameters and variables.

7.5 Numerical solutions

A wide range of software is now available for the numerical solution of the types of ODE system described above, some being intended for more general purposes, such as MATLAB, NAG routines or XPPAUT, and some being designed specifically for biological applications, for example Berkeley Madonna, CellDesigner, COPASI or the Systems Biology Toolbox for MATLAB. In this chapter, we employ MATLAB v. 7.8 (The MathWorks, Inc.) to obtain the time-dependent solutions and XPPAUT 5.91 for steady-state solutions.

Figure 7.4 illustrates a computational simulation of the model of the staphylococcal *agr* system using the default initial conditions (low levels of all the Agr proteins exist) and parameters from Jabbari *et al.*[58], with the exception that the ratio of basal transcription to quorum sensing-induced transcription is taken to be $\varepsilon = 0.01$. It should be noted that, for the reasons discussed above, the solutions provided here are those of the nondimensional model which, in the interests of brevity, is not recorded here (full details can, however, be found in Jabbari *et al.*[58]); the nondimensional model corresponds qualitatively to the model described in this chapter. Figure 7.5 allows us to see how a population of cells would shift from a downregulated state to an upregulated and *agr*-active one, see (xii) where P (the proportion of *agr* upregulated cells)

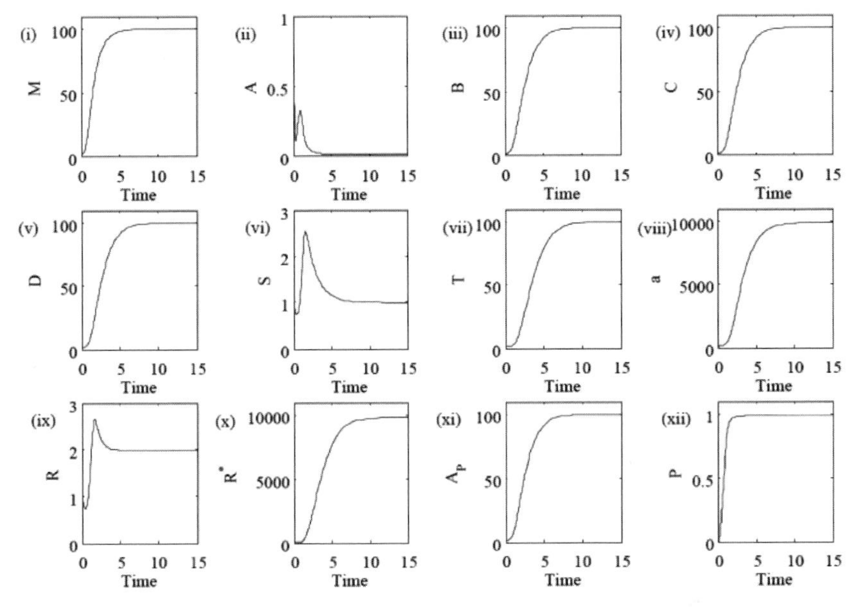

Fig. 7.4: We solve the equations given in Fig. 7.3 numerically and display the results here. We use the initial conditions and parameter scalings from Jabbari *et al.*[58]. Each graph corresponds to a different component of the agr system (see Table 7.3). The increase in AIP concentration in (viii) causes an increase in activator levels seen in (xi) which enforces the proportion of upregulated cells to approach unity in (xii).

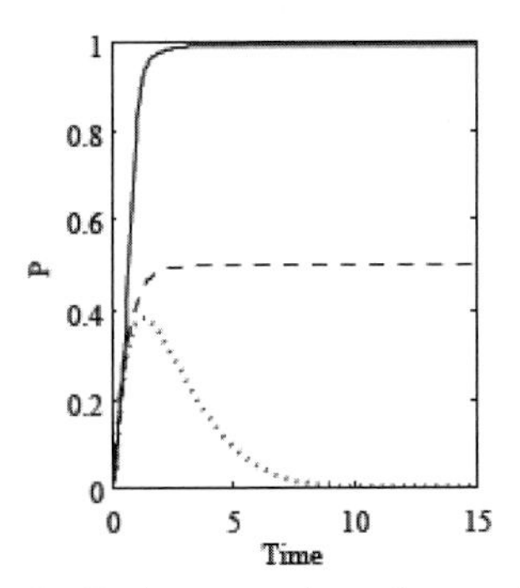

Fig. 7.5: Numerical solution for P, the proportion of upregulated cells, using the model given in Fig. 7.3 with initial conditions and parameter scalings from Jabbari *et al.*[58], subject to various mutations. Solid line: the full (wild-type) model; dashed line: phosphorylated AgrA-induced transcription from P2 is removed; dotted line: basal transcription is also removed from the model (the solution when only basal transcription is removed is indistinguishable from the wild-type model). In order for full upregulation of the cells to be achieved, QS-induced transcription must be in operation.

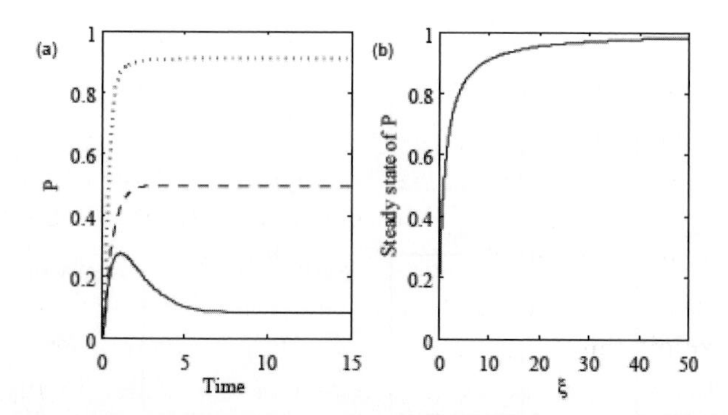

Fig. 7.6: (a) Numerical solution to P, the proportion of upregulated cells, using the model given in 7.3 with feedback into *agrA* and *agrC* transcription removed and replaced with a basal transcription rate, ξ. Solid line: $\xi = 0.1$; dashed line: $\xi = 1$; dotted line: $\xi = 10$; all other parameters are taken from the default set in Jabbari *et al.*[58]. It is evident that a low basal rate of transcription for *agrA* and *agrC* will prevent the majority of cells from reaching an upregulated state; this can be overcome for sufficiently high ξ. (b) The solution curve for P with varying ξ: the proportion of upregulated cells increases with ξ.

approaches 1. In terms of *S. aureus* this relates to the bacteria becoming virulent while in *C. acetobutylicum* this would be likely to relate to the decision to assign a proportion of cells to sporulation. Gene regulation models do not only permit simulations of what has been observed experimentally (as in Fig. 7.4), they also allow investigations into the various mechanisms involved in the network and, furthermore, can be manipulated to represent alterations to the genetic circuitry (for instance a gene mutation or overexpression) for the formulation of experimentally-testable hypotheses. For example, in Fig. 7.6 we demonstrate what the model predicts would occur if either basal (arrow I on the network diagram of Fig. 7.2(a)) and/or quorum sensing-induced (arrow II) transcription were to be removed from the model. We see that for full upregulation of a population, basal transcription is not required (though there must be at least a low level of Agr proteins present at the outset, i.e. that have not yet been lost through degradation or dilution post-mutation, for upregulation to occur), while, perhaps more surprisingly, basal transcription alone is sufficient to induce upregulation of roughly half of the cells (this proportion can be reduced by increasing the AIP loss rate). In order to render this proportion negligible under the parameter set used here, both basal and quorum sensing-induced transcription must be removed (dotted line of Fig. 7.6).

As discussed in Section 7.2, it is intriguing that the *agr* circuitry in *C. acetobutylicum* does not follow the same format as *S. aureus*: of the quorum sensing subloops, only that concerned with signal production is thought to exist in the former (Fig. 7.2(a) with arrows I and III). Mathematical models can be readily adapted to fit alternative scenarios and we exploit this to investigate why *C. acetobutylicum* may use a modified quorum sensing system to that of *S. aureus*. In Fig. 7.6 we provide the numerical solutions for the proportion of upregulated cells with the *agrCA* feedback loops removed, i.e. *agrCA* is transcribed only at a constant basal level.

We see that, unless this basal rate is increased significantly in comparison to that of *agrBD*, the proportion of *agr*-active cells is severely reduced, consistent with investigations in Jabbari *et al.*[58] which indicate that, in *S. aureus*, of the four quorum sensing subloops, that leading to induction of *agrA* transcription is the limiting factor in achieving upregulation of the full operon. In *S. aureus*, the induction of the *agr* operon is required for the bacteria to produce certain virulence factors required for infection, and coordinated behaviour by the entire population is likely to maximise the likelihood of a successful (for the bacteria) infection. The situation with *C. acetobutylicum*, where the *agr* operon appears to be linked with sporulation (see Section 7.2), is somewhat different: sporulation is required when the bacteria are subject to harsh conditions in which vegetative growth is unsustainable. In general, the entire population does not undergo sporulation: a number of vegetative cells will remain, presumably in case the conditions change and vegetative growth becomes the preferred state once again (the process of sporulation is highly time- and energy-consuming and cannot be reversed until after the spore is formed). Our modelling suggests that the proportion of *agr*-active cells (which is roughly equivalent to the proportion undertaking sporulation) can be adjusted according to the transcription level of *agrCA* in *C. acetobutylicum*. Thus, we might speculate that *C. acetobutylicum* uses this modified *agr* circuitry so that the cells can efficiently detect their population number (using the dotted feedback loop in Fig. 7.2(b)) in order to determine the desirability of sporulation, whilst sending only a portion of the population to this fate; this portion should survive until the conditions return to being suitable for vegetative growth and be sufficient in number to restore a healthy population of vegetative cells.

7.6 The relevance of modelling the *agr* operon to clostridial systems biology

If *C. acetobutylicum* is to be exploited for its butanol-producing capabilities, it is vital that the genetic mechanisms governing this and other related processes (such as sporulation) be understood as fully as possible. The careful application of gene-regulation models is a useful tool for such purposes because, as outlined above, they can allow investigations to be undertaken into the mechanisms involved, in particular by encapsulating the proposed mechanisms in a way that makes the hypotheses and assumptions involved completely clean cut and explicit; indeed the process of model formulation itself often brings to light hidden assumptions that might otherwise remain, at best, merely implicit. Importantly, they are particularly suitable for generating hypotheses upon synthetic-biology strategies: alterations to gene expression or gene mutations can often be represented in a gene regulation model simply by changing the parameter set. Thus, once the role of *agr* in *C. acetobutylicum* (and similarly in many other Gram-positive bacteria) has been elucidated, creating model-based hypotheses regarding synthetic-biology strategies can be easily facilitated. The most promising hypotheses can then be tested experimentally, thus reducing the amount of expensive and time-consuming work to be conducted in the laboratory.

Processes within a bacterial cell typically involve multiple gene regulation networks interacting with each other. Models such as that of the *agr* operon presented in this chapter can be combined with those of connecting networks with relative ease, thereby providing building blocks for more complicated models as well as being of interest in their own right. For instance, it would be worthwhile to analyse a model combining the *agr* operon with a model of the sporulation initiation network in *C. acetobutylicum* (in a similar vein to work previously carried out for *B. subtilis*[60,61]) in order to study the influence of quorum sensing upon sporulation. Similarly, if a point of contact can be found between the *agr* quorum sensing circuitry and solvent production in *C. acetobutylicum* it would be enlightening to investigate the effect of population density upon the yield of valuable products such as butanol and ethanol.

Acknowledgement

The authors gratefully acknowledge the support of the BBSRC for this work, which was undertaken as part of the SysMO project COSMIC (Systems Biology of *C. acetobutylicum* — a possible answer to dwindling crude oil reserves) forming part of the SysMO (Systems Biology of Microorganisms) initiative, and thank the other members of the consortium for helpful discussions. J. R. King also thanks the Royal Society and Wolfson Foundation for funding. Sara Jabbari was supported by a biomedical informatics fellowship from the MRC.

References

1. Williams P. *et al.*, *Phil. Trans. R. Soc. B.* 1119 (2007).
2. Atkinson S. and Williams P., *J. R. Soc. Interface* 959 (2009).
3. Sturme M. H *et al.*, *Antonie Van Leeuwenhoek* 233 (2002).
4. M. Pottathil and B. A. Lazazzera, *Front. Biosci.* 32 (2003).
5. Novick R. P. and Geisinger E., *Annu. Rev. Genet.* 541 (2008).
6. Auchtung J. M. *et al.*, *J. Bacteriol.* 5273 (2006).
7. Martin B. *et al.*, *Trends Microbiol.* 339 (2006).
8. Fontaine L. *et al.*, *J. Bacteriol.* 1444 (2010).
9. Shi K. *et al.*, *P. Natl. Acad. Sci. USA* 18596 (2005).
10. Dunny G. M., *Philos. Trans. R. Soc. Lond. B* 1185 (2007).
11. van Belkum M. J. *et al.*, *Microbiol.* 3660 (2007).
12. Fontaine L. *et al.*, J. *Bacteriol.* 7195 (2007).
13. Nakayama J. *et al.*, *Mol. Microbiol.* 145 (2001).
14. Slamti L. and Lereclus D., *EMBO J.* 4550 (2002).
15. Autret N. *et al.*, *Infect Immun.* 4463 (2003).
16. Riedel C. U., *et al.*, *Mol. Microbiol.* 1177 (2009).
17. Ohtani K., *et al.*, *J. Bacteriol.* 3919 (2009).
18. Senadheera D. and Cvitkovitch D. G., *Adv. Exp. Med. Biol.* 178 (2008).
19. Queck S. Y. *et al.*, *Mol. Cell.* 150 (2008).

20. Nakayama J. *et al.*, *J. Bacteriol.* 8321 (2006).
21. Qin X *et al.*, *J. Bacteriol.* 3372 (2001).
22. Sturme M. H. *et al.*, *J. Bacteriol.* 5224 (2005).
23. Rieu A. *et al.*, *Appl. Environ. Microbiol.* 6125 (2007).
24. Vidal J. E. *et al.*, *PLoS One* e6232 (2009).
25. Okumura K. *et al.*, *J. Bacteriol.* 7719 (2008).
26. Sebaihia M. *et al.*, *J. Genome Res.* 1082 (2007).
27. Wuster A. and Babu M. M., *J. Bacteriol.* 743 (2008).
28. Paredes C. J. *et al.*, *Nucleic Acids Res.* 1973 (2004).
29. Alsaker K. and Papoutsakis E. T., *J. Bacteriol.* 7103 (2005).
30. Errington J., *Nat. Rev. Microbiol.* 117 (2003).
31. Dürre P. and Hollergschwandner C., *Anaerobe* 69 (2004).
32. James S. *et al.*, *J. Mol. Biol.* 1127 (2000).
33. Dockery J. D. and Keener J. P., *Bull. Math. Biol.* 95 (2001).
34. Ward J. P. *et al.*, *IMA J. Math. Appl. Med.* 263 (2001).
35. Nilsson P. *et al.*, *J. Mol. Biol.*, 631 (2001).
36. Fagerlind M. G. *et al.*, *J. Mol. Microbiol. Biotechnol.* 88 (2003).
37. Anguige K. *et al.*, *Math. Biosci.* 39 (2004).
38. Haseltine E. L. and Arnold F. H., *Appl. Environ. Microbiol.* 437 (2008).
39. Kuttler C. and Hense B. A., *J. Theor. Biol.* 167 (2008).
40. Anguige K. *et al.*, *J. Math. Biol.* 557 (2005).
41. Anguige K. *et al.*, *Math. Biosci.* 240 (2006).
42. Fagerlind M. G. *et al.*, *BioSystems* 201 (2005).
43. Ward J. P. *et al.*, *Math. Med. Biol.* 169 (2004).
44. Koerber A. J. *et al.*, *Bull. Math. Biol.* 239 (2002).
45. Chopp D. L. *et al.*, *Bull. Math. Biol.* 1053 (2003).
46. Ward J. P. *et al.*, *J. Math. Biol.* 23 (2003).
47. Smith C. *et al.*, *J. Mol. Biol.* 1290 (2008).
48. Müller J. *et al.*, *J. Math. Biol.* 672 (2006).
49. Hense B. A. *et al.*, *Nat. Rev. Mirobiol.* 230 (2007).
50. Koerber A. J. *et al.*, *J. Math. Biol.* 440 (2005).
51. Goryachev A. B. *et al.*, *PLoS Comp. Biol.* 265 (2005).
52. Goryachev A. B. *et al.*, *BioSystems*, 178 (2006).
53. Li J. *et al.*, *Mol. Sys. Biol.* 1 (2006).
54. Hong D. *et al.*, *J. Theor. Biol.* 726 (2007).
55. Müller J. *et al.*, *BioSystems* 76 (2008).
56. Gustafsson E. *et al.*, *J. Mol. Microbiol Biotechnol.* 232 (2004).
57. Jabbari S., Ph.D. Thesis, University of Nottingham, UK (2007).
58. Jabbari S. *et al.*, *J. Math. Biol.* DOI:10.1007/s00285 (2009).
59. Jabbari S. *et al.*, *Math. Biosci.* DOI:10.1016/j.mbs.2010.03.001 (2010).
60. Bischofs I. B. *et al.*, *P Natl. Acad. Sci. USA* 6459 (2009).
61. Jabbari S. *et al.*, *Bull. Math. Biol.*, DOI: 10.1007/s11538-010-9530-7 (2010).

CHAPTER 8

COMPARATIVE GENOMIC ANALYSIS OF THE CENTRAL METABOLISM OF THE SOLVENTOGENIC SPECIES *CLOSTRIDIUM ACETOBUTYLICUM* ATCC 824 AND *CLOSTRIDIUM BEIJERINCKII* NCIMB 8052

M. A. J. SIEMERINK, K. SCHWARZ, W. KUIT, S. W. M. KENGEN

Laboratory of Microbiology, Wageningen University
Dreijenplein 10, 6703HB
Wageningen, NL

C. GRIMMLER, A. EHRENREICH

Department of Microbiology, Technical University München
Emil-Ramann-Strasse 4
85354 Freising, DE

Solvent-producing clostridia have attracted renewed scientific interest because of their potential to produce biofuels. *Clostridium acetobutylicum* ATCC824 and *Clostridium beijerinckii* NCIMB 8052 are the best-studied solventogenic species, and the availability of their entire genome sequences offered the possibility to compare both solventogenic species at the gene level. General genome features as well as COG (clusters of orthologous groups of proteins) categorizations were summarized. All genes coding for the various proteins of the main catabolic pathway were discussed and compared, including paralogous and orthologous genes. Whenever possible, the most likely catabolic gene was identified, based on available experimental data. Overall, for most enzymatic steps comparable genes were found in both genomes, although *C. beijerinckii* in general harbors more paralogs, resulting also in a relatively large genome size. Essential differences were found as well. *C. acetobutylicum* contains a pyruvate decarboxylase, while *C. beijerinckii* contains an Rnf cluster and a trimeric bifurcating hydrogenase. Moreover, in *C. acetobutylicum* most of the solventogenic genes are located on the pSOL1 megaplasmid, which is absent in *C. beijerinckii*. The physiological background of the observed similarities and differences was discussed.

8.1 Introduction

Since the beginning of the 20th century, many clostridial strains have been isolated because of their capacity to produce acetone and butanol. These solvents, produced by starch- or molasses-based fermentations, were important building blocks for the industrial production of various chemicals. During the 1950s the production of solvents by fermentation became, however, outcompeted by cheaper petroleum-based processes[1]. The current interest in the use of renewable resources as feedstock for the production of fuels has resulted in a revival of the research on the fermentation route as a possible source of solvents[2,3]. The numerous clostridial strains that were isolated at the time, have recently been reassigned into four species by 16SrRNA analysis and DNA fingerprinting, *viz. C. acetobutylicum, C. beijerinckii, C. saccharoper-butyl-acetonicum*, and *C. saccharobutylicum*[4]. Of these clostridia, the genome of the type strain C. *acetobutylicum* ATCC 824 as well as the *C. beijerinckii* strain NCIMB 8052 has been sequenced. Although both species are well-known for their capacity to produce solvents, essential differences exist in their physiology, thus making a comparative genome analysis attractive.

C. acetobutylicum has been studied the most[5,6]. It is able to utilize a wide variety of carbohydrates, e.g. glucose, maltose, sucrose, lactose, and cellobiose. In addition, it can also hydrolyze gelatin, produce riboflavin, and is sensitive to rifampicin[4]. *C. beijerinckii* has a similar substrate spectrum, but can in addition also grow on several sugar alcohols. However, it cannot hydrolyze gelatin or produce riboflavin and most strains are not sensitive to rifampicin[4]. In contrast to *C. aceto-butylicum*, some *C. beijerinckii* strains have the capacity to produce isopropanol[7]. It has been discussed that the differences in substrate utilization reflect the original isolation conditions of *C. acetobutylicum* and *C. beijerinckii*, being starch and molasses, respectively[8]. With respect to their solvent-forming ability, *C. acetobutylicum* forms the highest solvent levels under low pH conditions (below pH 5)[9], whereas *C. beijerinckii* is able to produce significant amounts already at pH 7[10].

Furthermore, in chemostat culture, *C. beijerinckii* ultimately always gives rise to a population dominated by degenerated mutant(s) that are not capable of solvent production, whilst *C. acetobutylicum* under similar culture conditions sustains solventogenesis[11,12]. Although less frequent, degeneration of *C. acetobutylicum* can occur under specific conditions in a chemostat as well, but appears to result from loss of the megaplasmid pSOL1, which carries several essential solvent-forming genes[13].

The availability of the genomes of *C. acetobutylicum* ATCC 824[14] and *C. beijerinckii* NCIMB 8052 (http://genome.ornl.gov/microbial/cbei/) now offers the opportunity to compare both solventogenic species at the gene level. This chapter aims to provide such a comparative genomic analysis, which can help to explain the observed phenotypic differences, but may also reveal general features of solvent production and lead to a better understanding of both solventogenic species. First some general features will be described followed by paragraphs focusing in more detail on the central metabolic pathways as well as energy generation and the disposal of reductant. Genome sequences of clostridia that are less studied or that are not solventogenic will not be discussed here. Complete genome sequences are also available for the pathogenic strains of *C. botulinum* and *C. difficile* and although these species are capable of producing some butanol, they are for obvious reasons not of interest for biofuel production, and will also not be considered here.

8.2 General genome features

The strains that have been used for genome comparison are *C. acetobutylicum* ATCC 824 (DSM-792), which is the type-strain, and *C. beijerinckii* NCIMB 8052 (DSM-1739). If not indicated otherwise, we refer to these two strains in the remaining part of this chapter. The NCIMB strain has been shown by DNA-DNA reassociation analysis (73–83% similarity) to differ from the actual type-strain (VPI 5481, ATCC 25752, DSM 791), but to remain within the species range[4,15]. Based on a partial 16S rRNA gene sequence, *C. acetobutylicum* and *C. beijerinckii* are 90% identical[15]. The entire genome sequences of

C. acetobutylicum and *C. beijerinckii* are deposited under GenBank accession numbers AE001437 and CP000721, respectively. For genome comparisons information used is available at the IMG-JGI website (http://img.jgi.doe.gov/cgi-bin/w/main.cgi). A detailed comparison at amino acid level between all proteins of *C. acetobutylicum* and *C. beijerinckii* was done by a computerized all-to-all alignment using ClustalX[16]. (*C. acetobutylicum* versus *C. acetobutylicum*, and *C. acetobutylicum* versus *C. beijerinckii*.) A cut-off value of 40 percent identity was chosen, meaning paralogs and orthologs with an identity of less than 40% were discarded. This, however, does not exclude the fact that some of the low-identity orthologs, or even phylogenetically non-related proteins, might be responsible for the same reaction.

As described above, both species are phenotypically rather distinct. This difference can also be observed from their general genome features, as listed in Table 8.1. The most apparent difference is the larger genome size of *C. beijerinckii* (6 Mbp), not only in comparison to *C. acetobutylicum* (3.9 Mbp) but to all other completely sequenced

Table 8.1: General features of two solventogenic clostridia, *C. acetobutylicum* ATCC 824 and *C. beijerinckii* NCIMB 8052.

Species	*Clostridium acetobutylicum* ATCC 824		*Clostridium beijerinckii* NCIMB 8052	
	Number	%	Number	%
GC %		31		30
Number of bases	3,940,880		6,000,632	
CDS %		95.67		96.41
Genes	3,846		5,290	
Protein genes with funct. pred.*	3,101	77	3,796	72
Pseudogenes	0		80	1.5
RNA genes	174		190	
Total number of enzymes *	821	20	1,070	20
Orthologs *	3,619	90	4,815	91
Paralogs *	2,118	53	3,364	64
COGs *	2,808	70	3,839	73
CRISPR sequences	0		0	
Plasmid	pSOL1		-	
Number of bases	192,000		-	
Number of genes	176		-	
GC %		31	-	

* Including genes on plasmid.

clostridia, whose genomes sizes are ~4 Mbp (www.ncbi.nlm.nih.gov/genome/?term=clostridium) or lower. Another striking difference is the presence of a 192 kb-megaplasmid in *C. acetobutylicum* ATCC824 (pSOL1). Whereas most clostridial plasmids remain cryptic, pSOL1 is functional and harbors several of the genes (*aad, ctfA, ctfB, adc*) necessary for acetone and butanol formation[13]. *C. beijerinckii* has no indigenous plasmids, but it supports replication of a variety of vectors based on plasmids from *C. butyricum*, *Enterococcus faecalis*, and *Streptococcus cremoris*, which makes it more amenable to genetic manipulation[17]. *C. beijerinckii* lacks the active restriction system that is found in the *C. acetobutylicum* ATCC 824[18].

Both species are devoid of CRISPR sequences, which suggests that other ways of acquiring immunity against virus attacks are employed[19]. About three-quarters of the sequenced (draft and finished) clostridial genomes, however, do contain CRISPRs.

The genome of *C. beijerinckii* shows a comparable number of coding sequences, genes, and COGs (clusters of orthologous groups of proteins). The number of paralogs is, however, substantially higher, which suggests that the higher genome size is partly caused by extensive gene duplication events[20]. For instance, for transketolase, four paralogs exist in *C. beijerinckii*, in addition to three transketolase subunits A and B. *C. acetobutylicum* has only one transketolase gene. *C. beijerinckii* has nine paralogs for a PTS lactose/cellobiose IIB subunit, whereas *C. acetobutylicum* contains only one. For a small acid-soluble spore protein 11 paralogs are present in *C. beijerinckii*, while *C. acetobutylicum* has none. Some of these gene duplications will be discussed in more detail in the following paragraphs. The reason for the relatively higher percentage of paralogs is not known, although the PTS homologs may be a reflection of a broader substrate spectrum, as discussed below.

Gene order is generally poorly conserved in bacteria, even among closely related genomes[21]. Also, a genomic dot plot comparison of *C. acetobutylicum* with *C. beijerinckii* (not shown) revealed only few regions of colinearity. Thus, despite the relatively close evolutionary distance between *C. acetobutylicum* and *C. beijerinckii*, hardly any memory of the ancestral gene order has been retained. The genomes were compared more closely by analyzing the number of genes in specific

COG categories[22]. The COG database is a tool for the phylogenetic classification of proteins encoded in complete genomes, and which allows the prediction of gene functions. The number of genes in certain COG categories appears to differ markedly between both species (Fig. 8.1). *C. beijerinckii* shows considerably higher numbers for the categories transcription (Category K; 74% higher), signal transduction (Category T; 82% higher), carbohydrate transport and metabolism (Category G; 91% higher), and energy production and conversion (Category C; 91% higher). The increase Category K (transcription) found for *C. beijerinckii*, mainly concerns transcriptional regulators, whose function is often not clear, but is probably related to sugar metabolism (AraC, GntR, RpiR, GutR) and stress response (TetR, MarR, MerR, HxlR). The higher protein numbers involved in signal transduction (Category T) of *C. beijerinckii* compared to *C. acetobutylicum* primarily involves response regulators with a CheY-like domain (83 and 40 resp.), histidine kinases (82 and 34, resp.) and methyl accepting chemotaxis proteins (61 and 36, resp.). Within Category G (carbohydrate transport and metabolism), differences relate to various PTS system components as well as ABC-transporter components, and some pentose phosphate pathway enzymes. Within Category C (energy production and conversion), the main differences concern aldehyde- and alcohol dehydrogenases and various other redox proteins (NAD(P)-dependent and flavodoxin- or ferredoxin-related proteins). *C. beijerinckii* has a higher number of nitroreductases compared to *C. acetobutylicum* (24 and 11, respectively), which are likely to be involved in resistance to environmental stresses like antibiotics, toxic compounds or heavy metals[23].

This role in stress response is supported by their proximity to various regulators of the type MarR, TetR, MerR, and HxlR, as already mentioned for Category K. Overall, these data suggest that *C. beijerinckii* is better equipped to deal with various environmental stresses, and also has a much broader capacity for substrate utilization[4], involving more transport systems (ABC-type as well as PTS systems) for different carbohydrates, and herewith connected signal transduction (two-component systems) and transcriptional regulation proteins. The broader substrate range of *C. beijerinckii*, to a lesser extent, also holds

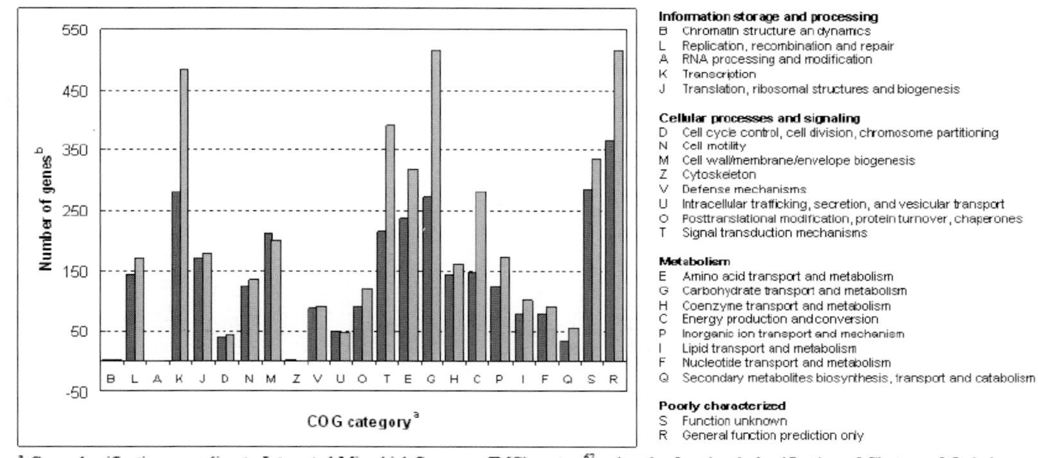

[a] Gene classification according to Integrated Microbial Genomes (IMG) system[67] using the functional classification of Clusters of Orthologous Groups of proteins (COG).[22] The distribution of genes over COG categories is not non-redundant, meaning that certain genes may be present in more than one category.
[b] Number of protein-coding genes in each category, including plasmid genes, but without pseudogenes.

Fig. 8.1: Functional categories of predicted ORFs in the genomes of *C. acetobutylicum* ATCC 824 and *C. beijerinckii* NCIMB 8052.

for uptake of other growth factors, like amino acids (Category E) and inorganic ions (Category P). General housekeeping genes do not show large differences between both species, for instance the genes involved in translation, cell division, and cell wall biogenesis (Category J, D, and M). One should realize, however, that for ~40% of the identified genes in both genomes a function has not been assigned.

Another obvious difference is that *C. beijerinckii* has a substantially higher number of transposases and derivatives thereof (39) than *C. acetobutylicum* (8). This was not evident from the COG comparison, because almost half (18) of the transposases of *C. beijerinckii* have not been assigned to a COG. The high transposase number in *C. beijerinckii* suggest that its genome is far more prone to rearrangements than that of *C. acetobutylicum*, and it has been discussed that insertion sequences (IS) might play a role in the rapid degeneration of solventogenesis observed for *C. beijerinckii*[24].

8.3 The central catabolic pathways

This paragraph focuses on the enzymes involved in the glycolysis (Embden–Meyerhof pathway) and subsequent pathways leading to the main end-products of the acidogenic and solventogenic metabolism, *viz.* acetate, butyrate, lactate, acetone, acetoin, ethanol, butanol, carbon dioxide, and H_2 (Fig. 8.2 and Table 8.2). The metabolic network as depicted in Fig. 8.2 and Table 8.2 was based on the Kyoto Encyclopedia of Genes and Genomes (http://www.genome.jp/kegg/).

In *C. acetobutylicum* ATCC 824, 34 reactions are required to compose the glycolysis and subsequent pathways (Fig. 8.2). For these 34 reactions, 53 genes (including paralogs) have been annotated in KEGG, and for most reactions it has been experimentally proven which of the paralogs is actually coding for a specific enzyme activity. For the 53 genes encoding central metabolism enzymes of *C. acetobutylicum*, 59 orthologs could be retrieved from the *C. beijerinckii* genome. Despite the higher number of orthologs in *C. beijerinckii*, for six of the *C. acetobutylicum* genes (i.e. #1, CA_C0570; #7, CA_C2018; #15, CA_P0025; #23, CA_C2967; #26, CA_C0028; and #30, CA_C3375) no close ortholog was detected.

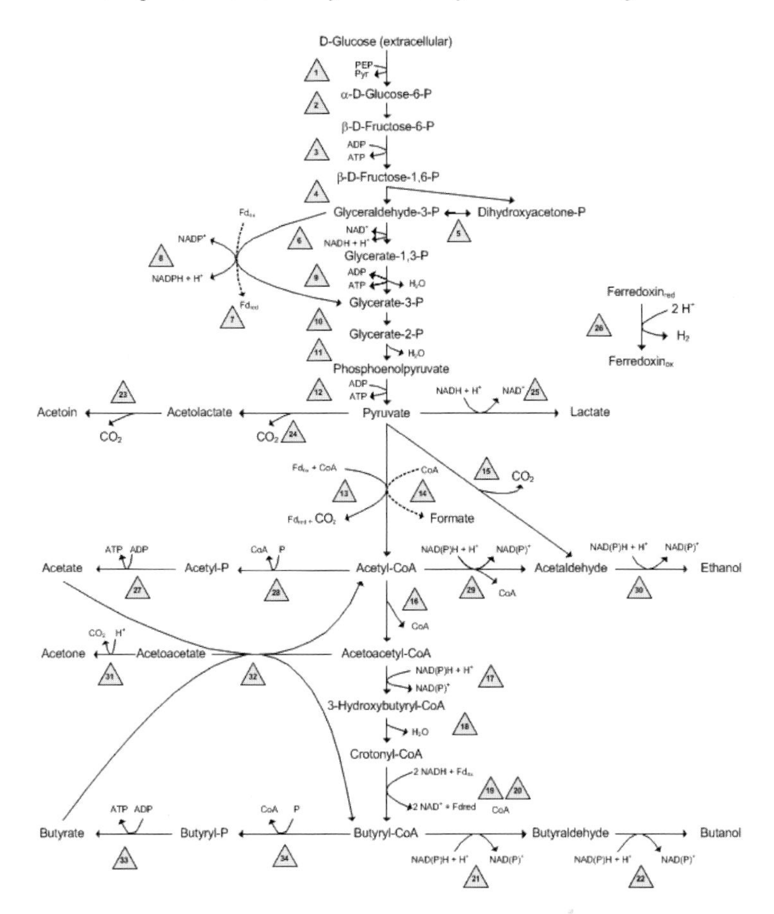

Fig. 8.2: Central glycolytic metabolism of two solventogenic clostridia, *C. acetobutylicum* ATCC 824 and *C. beijerinckii* NCIMB 8052. Numbers in triangles refer to the reaction numbering in Table 8.2.

In the following paragraphs, the main enzymatic steps of the acidogenic and solventogenic pathways of both species and their encoding genes will be discussed in more detail.

8.3.1 Glucose uptake

Glucose is taken up via phosphoenolpyruvate (PEP)-phosphotransferase systems (PTS). There is no indication of ABC-type transport systems or secondary transport systems for glucose in *C. acetobutylicum*. Based on

Table 8.2: Enzymes and corresponding genes involved in the central metabolism of *C. acetobutylicum* and *C. beijerinckii.* The Table shows all candidate enzymes of *C. acetobutylicum* indicated in Figure 2, their parakigs and the respective orthologs in *C. beijerinckii* on amino acid (AA) level. The cut-off value is 40% identity (ID). Numbers in superscript indicate gene positions within gene clusters. COG numbering as indicated in IMG. The most likely enzyme candidates within *C. acetobutylicum* and *C. beijerinckii,* based on gene neighborhood, homology, microarray data (differential expression) and literature information are shown in bold.

#	EC	COG	Name	Locus-tag	AA	Locus-tag	AA	Id (%)	Ref.
				C. acetobutylicum		*C. beijerinckii*			
1	[2.7.1.69]	1263;	PTS enzyme II, ABC components	**CA_C0570 *(glcG)***	665	*nd*			[26,28]
		1264;	BC components	CA_C3425^{1} *(glvC)*	518	CBEI_4977^{2}	513	80	
		2190	A component	CA_C3427^{2}	165	CBEI_4982^{3}	158	41	[28]
			BC components	CA_C1353^{1}	488	CBEI_4532^{1}	478	41	[27,28]
			A component	CA_C1354^{2}	159	CBEI_4533^{2}	165	53	[28]
						CBEI_4706	158	46	
			A component	CA_C2995	157	CBEI_3273	634	41	[28]
2	[5.3.1.9]	0166	Glucose-6-phosphate isomerase (PGI)	**CA_C2680 *(pgi)***	450	**CBEI_0341**	449	72	[28,34,68-70]
3	[2.7.1.11]	0205	6-Phosphofructokinase (PFK)	**CA_C0517^{1} *(pfkA)***	319	**CBEI_4852^{2}**	318	76	[28,34,68,69,72]
4	[4.1.2.13]	3588	Fruct.-1,6-bisphosphate aldolase (FBA)	CA_P0064 *(alf)*	295	CBEI_3039	295	79	[28]
				CA_C0827 *(fba)*	287	**CBEI_1903**	288	81	[27,28,34,68-70]
5	[5.3.1.1]	0149	Triosephosphate isomerase (TPI)	**CA_C0711^{3} *(tpiA)***	248	**CBEI_0599^{3}**	248	73	[28,34.68-73]
6	[1.2.1.12]	0057	GAP dehydrogenase (GAPDH)	**CA_C0709^{1} *(gapC)***	334	**CBEI_0597^{1}**	333	75	[28,34,68-71]
7	[1.2.7.5]	2414	Aldehyde:ferredoxin oxidoreductase	CA_C2018	684	nd			[27,34]
8	[1.2.1.9]	1012	NADP-dep. GAP dehydrog. (GAPN)	**CA_C3657^{2} *(gapN)***	482	**CBEI_2572**	486	65	[34,69,70,74]
						CBEI_2282	488	63	

Table 8.2: (*Continued*)

9	[2.7.2.3]	0126	Phosphoglycerate kinase (PGK)	**CA_C0710[2] *(pgk)***	397	**CBEI_0598[2]**	391	79	[28,68,71]
10	[5.4.2.1]	0588	Phosphoglycerate mutase (PGM)	**CA_C0712[4] *(pgm)***	510	**CBEI_0600[4]**	512	71	[28,34,68-71]
				CA_C2741 *(gpmA)*	243	CBEI_1759	232	70	
				CA_C0167	228	CBEI_3922	237	58	[72]
				CA_C3021	219	CBEI_1719	203	50	
11	[4.2.1.11]	0148	Enolase (ENO)	**CA_C0713[5] *(eno)***	431	**CBEI_0602[4]**	430	76	[28,34,68,70,72]
12	[2.7.1.40]	0469	Pyruvate kinase (PYK)	**CA_C0518[2] *(pykA)***	473	**CBEI_4851[1]**	476	64	[28,34,68-71]
						CBEI_0485	472	55	
				CA_C1036 *(pykA)*	472	CBEI_0485	472	60	[28,68,70]
						CBEI_4851[1]	473	57	
13	[1.2.7.1]	0674; 1013	Pyruvate:Fd oxidoreductase (PFOR)	**CA_C2229 *(pfor)***	1171	**CBEI_4318**	1169	72	[34,69,70,75]
						CBEI_1853	1171	66	
						CBEI_1458	1172	64	
				CA_C2499	1173	CBEI_1458	1172	73	[69]
						CBEI_4318	1169	63	
						CBEI_1853	1171	62	
14	[2.3.1.54]	1882	Pyruvate: formate lyase	CA_C0980	743	CBEI_1009_	743	72	[37,38]
						CBEI_1011	743	72	
15	[4.1.1.1]	3961	Pyruvate decarboxylase	**CA_P0025**	554	nd			[39]
16	[2.3.1.9]	0183	Thiolase (THL)	**CA_C2873 *(thlA)***	392	**CBEI_3630**	392	90	[27,28,34,68-71]
						CBEI_0411	393	74	
				CA_P0078 *(thiL)*	392	CBEI_3630	392	78	[28,34,71]
						CBEI_0411	393	76	
17	[1.1.1.157]	1250	3-HO-butyryl-CoA dehydrog. (HBD)	**CA_C2708[1] *(hbd)***	282	**CBEI_0325[5]**	282	79	[28,34,68-71,76]
18	[4.2.1.55]	1024	Crotonase (CRT)	**CA_C2712[5] *(crt)***	261	**CBEI_0321[1]**	261	69	[28, 34,68-72,76]

Table. 8.2: (*Continued*)

						CBEI_2034[1]	261	69	
						CBEI_4544[2]	258	57	
19	[1.3.99.2]	1960	Butyryl-CoA dehydrogenase (BCD)	**CA_C2711[4] *(bcd)***	379	**CBEI_0322[2]**	379	79	[28, 34,68-71,74,76]
						CBEI_2035[2]	379	79	
						CBEI_2883	379	65	
						CBEI_4542[1]	379	60	
20		2025	Electron transfer flavoprotein α (EtfA)	**CA_C2709[2] *(etfA)***	336	**CBEI_0324[4]**	335	71	[31,49,50]
						CBEI_2037		71	
		2086	Electron transfer flavoprotein β (EtfB)	**CA_C2710[3] *(etfB)***	259	**CBEI_0323[3]**	259	71	[31,49,50]
						CBEI_2036		71	
21	[1.2.1.10]	1012;	Alcohol-aldehyde dehydrog. (AADH)	**CA_P0035 *(adhE2)***	858	**CBEI_0305**	864	66	[27,28,34,69,71,74,76]
		1454		**CA_P0162[1] *(aad)***	862	**CBEI_0305**	862	66	[27,28,34,68-71,74,76]
22	[1.1.1.-]	1063;	Butanol dehydrogenase (BDH I,II)	**CA_C3299[2] *(bdhA)***	389	**CBEI_2421**	387	64	[27,28,34,68-72,74,76]
				CA_C3298[1] *(bdhB)*	390	**CBEI_2421**	387	60	[27,28,34,68-72,76]
23	[4.1.1.5]	3527	Acetolactate decarboxylase	**CA_C2967**	238	nd			
24	[2.2.1.6]	0028;	Acetolactate synthase AB	CA_C3169[2] *(ilvB)*	554	CBEI_0217[2]	557	54	[28,34,72]
		0440				CBEI_2646	560	40	
				CA_C3176[2] *(ilvN)*	165	CBEI_0212[1]	169	48	[28,68]
				CA_C3652[1] *(alsS)*	563	CBEI_2868	556	63	
25	[1.1.1.27]	0039	L-lactate dehydrogenase (LDH)	**CA_C0267 *(ldh)***	313	**CBEI_1014**	316	58	[34,69,72,74]
				CA_C3552	320	CBEI_4126	553	69	[28,34,70]
						CBEI_4972[1]	535	50	
26	[1.12.7.2]	4624	Hydrogenase (HYDA) - (iron only)	**CA_C0028 *(hydA)***	582	CBEI_3796	461	40	[27,28,68,71,72,77]
						CBEI_4110	567	35	[58]
		4624	Hydrogenase (HYDB) – (iron only)	CA_C3230 *(hydB)*	450	CBEI_1901	449	54	

Table. 8.2: (*Continued*)

	[1.12.99.6]	1740	Hydrogenase (HYD) - (nickel-iron)	**CA_P0141[1] *(mbhS)***	291	**CBEI_3013[2]**	291	70	[28]
		0374		**CA_P0142[2] *(mbhL)***	471	**CBEI_3012[1]**	463	62	[28]
27	[2.7.2.1]	0282	Acetate kinase (ACK)	**CA_C1743[2] *(askA))***	401	**CBEI_1165[2]**	400	59	[28,34,68,70,71,76]
	[3.6.1.7]	1254	Acylphosphatase (ACYP)	CA_C2830	91	CBEI_4233	90	52	
28	[2.3.1.8]	0280	Phosphotransacetylase (PTA)	**CA_C1742[1] *(pta)***	333	**CBEI_1164[1]**	333	68	[28, 34,68-71,76]
29	[1.2.1.10]	1012;	Alcohol-aldehyde dehydrog. (AADH)	**CA_P0035 *(adhE2)***	858	**CBEI_0305**	864	66	[27,28,34,70,71,74,76]
		1454		**CA_P0162[1] *(aad)***	862	**CBEI_0305**	864	66	[28, 34,68-71,74,76]
30	[1.1.1.1]	1063	Acetaldehyde dehydrogenase (ACDH)	CA_C3375	377	nd			
30	[1.2.1.10]	1012; 1454	Alcohol-aldehyde dehydrogenase (AADH)	**CA_P0035 *(adhE2)***	858	**CBEI_0305**	864	66	[27,28,34,70,71,74,76]
				CA_P0162[1] *(aad)*	862	**CBEI_0305**	864	66	[28, 34,68-71,74,76]
31	[4.1.1.4]	4689	Acetoacetate decarboxylase (AADC)	**CA_P0165[4] *(adc)***	244	**CBEI_3835[3]**	246	75	[27,28,34,68-74,76]
32	[2.8.3.9]	1788;	CoA transferase AB (CoAT AB)	**CA_P0163[2] *(ctfA)***	218	**CBEI_3833[1]**	217	73	[27,28,34,68-72,74,76]
		2057				CBEI_2654[2]	238	42	
				CA_P0164[3] *(ctfB)*	221	**CBEI_3834[2]**	221	69	[27,28,34,68-72,74,76]
						CBEI_2653[1]	217	51	
33	[2.7.2.7]	3426	Butyrate kinase (BK)	**CA_C3075[1] *(buk)***	355	**CBEI_0204[2]**	355	64	[28,34,68,69,71,74]
						CBEI_4006	356	65	
						CBEI_4609	356	64	
				CA_C1660	356	CBEI_0204[2]	355	62	[51]
						CBEI_4006	356	74	
						CBEI_4609	356	71	
34	[2.3.1.19]	0280	Phosphotransbutyrylase (PTB)	**CA_C3076[2] *(ptb)***	301	**CBEI_0203[1]**	302	71	[28,34,68-71,74,76]

the genome sequence, *C. acetobutylicum* contains 13 PTS systems. Six of these systems belong to the glucose-glucoside (Glc) family, which contains permeases that are specific for glucose, N-acetylglucosamine, or different α- and β-glucosides[25]. For one system (containing all three necessary A, B, and C subunits) within this family, encoded by *glcG* (CA_C0570), it has been proven that it transports glucose inside the cells. However, the possibility of other active glucose specific phosphotransferases encoded by one of the other five subunit paralogs, remains[26]. Differential expression of some other PTS system subunits has even been observed[27,28], suggesting that in glucose grown cultures, alternative uptake proteins are likely to be involved. Furthermore, it has been speculated that one of the small subunits with a IIA^{glc} domain (CA_C2995) which is thought to be involved in glucose uptake, is possibly involved in maltose uptake, since the putative *mal* operon of *C. acetobutylicum* does not encode this IIA domain[25]. Remarkably, no ortholog could be found for the PTS component IIABC encoded by *glcG* (CA_C0570) within *C. beijerinckii*, although at least two clusters coding for $IIABC^{glc}$ domains (CBEI_4982, CBEI_4983 and CBEI_4532, CBEI_4533) do exist. Additionally, CBEI_4983, part of one of the two $IIABC^{glc}$ clusters is not shown in Table 8.2, but encodes domain IIBC (518 AA) with an identity of 33%. Since this ortholog had an identity lower than 40%, it was omitted in Table 8.2.

8.3.2 Glycolysis

The PTS systems result in the formation of glucose-6-P, which subsequently follows the established glycolytic steps to fructose-6-P and fructose-bis-P. For the encoding genes of the involved enzymes, *viz.* glucose-6-P isomerase and phosphofructokinase, only one candidate exists in *C. acetobutylicum* and *C. beijerinckii*. In both species, the phosphofructokinase gene (CA_C0517, CBEI_4852) lies clustered with the pyruvate kinase (CA_C0518, CBEI_4851), which is commonly observed in many fermentative anaerobes and which supports their role in glycolysis. There are no indications of alternative ADP- or PPi-dependent phosphofructokinases in either species, as have been reported for certain thermophilic sugar fermentors (*Pyrococcus furiosus*,

Caldicellulosiruptor saccharolyticus, *Thermoanaerobacter tengcongensis*, *Thermotoga maritima*). For the aldolase, *C. acetobutylicum* and *C. beijerinckii* contain two candidates, a class I and a class II type, of which one resides on the megaplasmid for *C. acetobutylicum* (CA_P0064). The chromosomally encoded class II aldolase (CA_C0827) is expressed constitutively and is most likely responsible for the glycolytic activity[29]. Similarly, it is expected that the only class II ortholog (CBEI_1903) in *C. beijerinckii* fulfils the glycolytic role. The genes involved in the subsequent carbon-3 part of the glycolysis are clustered on the chromosome for *C. acetobutylicum* as well as *C. beijerinckii*, again supporting their concerted action in glycolysis. These concern the genes for triose-P-isomerase, glyceraldehyde-3-P dehydrogenase, phosphoglycerate kinase, phosphoglycerate mutase, and enolase with locus tags CA_C0709 - CA_C0713 and CBEI_0597 - CBEI_0602 for *C. acetobutylicum* and *C. beijerinckii*, respectively. The *gap*, *pgk* and *tpi* have been shown to constitute an operon in *C. acetobutylicum*[30]. For *C. beijerinckii* it has been shown that *gap*, *pgk, tpi*, and a transcriptional regulator gene (CBEI_0596) also form an operon[31]. The adjacent *pgm* and *eno* are not part of the operon of both species. The conversion of GAP by GAPDH and PGK constitutes the first oxidation step and the first site for substrate level phosphorylation. However, for the conversion of GAP, both species also contain a NADP-dependent *gapN* homolog (CA_C3657, CBEI_2572), which directly oxidizes GAP to 3-phosphoglycerate, without formation of adenosine triphosphate (ATP)[32]. Its role in the glycolysis is unclear, although it has been reported to be activated together with other glycolytic enzymes under conditions of oxygen stress in *C. acetobutylicum*[33]. A role of GAPN in providing NADPH for biosynthetic purposes is likely. In addition, *C. acetobutylicum* but not *C. beijerinckii*, harbors an aldehyde:ferredoxin oxidoreductase (CA_C2018). It should be noted that this enzyme has not yet been characterized and proven to actually catalyze GAP oxidation, although, differential expression of CA_C2018 has been observed in butanol challenged cultures of *C. acetobutylicum*[27,34]. For the phosphoglycerate mutase, three paralogs were found in both *C. acetobutylicum* and *C. beijerinckii* (CA_C0167, CA_C2741, CA_C3021, CBEI_3922, CBEI_1759, CBEI_1719) in addition to the phosphoglycerate mutase gene (CA_C0712, CBEI_0600), in the glycolytic cluster. The latter belongs to the 2,3-phosphoglycerate independent monophosphoglycerate mutases with a size of ~510 residues, whereas the paralogs are dependent on 2,3-diphosphoglycerate and consist of ~230 residues. *C. acetobutylicum* and *C. beijerinckii* each contain two pyruvate kinase genes, of which one (CA_C0518, CBEI_4851) forms an operon with the phosphofructokinase gene[31] and most likely operates in the glycolysis. The role of the other pyruvate kinase (CA_C1036, CBEI_0485) is not known. The pyruvate kinase step constitutes the second substrate level phosphorylation site.

8.3.3 Gluconeogenesis

Most enzymes in glycolysis are reversible and can therefore also operate in the anabolic gluconeogenic direction, except for the fructose-1,6-bisphosphatase and the pyruvate kinase. Both species, *C. acetobutylicum* and *C. beijerinckii* contain two genes (CA_C1088, CA_C1572, CBEI_2467, CBEI_4541) annotated as fructose-1,6-bisphosphatase. The smaller protein, CA_C1088 consisting of 324 amino acids has no detectable orthologs, while CA_C1572 consisting of 665 amino acids has two orthologs, *viz.* CBEI_2467 (663 AA) and CBEI_4541 (653 AA) with an identity of 71% and 66%, respectively. Furthermore, CBEI_4541 is also encoded in close proximity to two other enzymes of the central metabolism, *viz.* 3-hydroxybutyryl-CoA dehydratase and butyryl-CoA dehydrogenase.

For the interconversion of PEP and pyruvate, two alternatives are known aside from the classical pyruvate kinase reaction, *viz.* pyruvate phosphate dikinase (PPDK) and PEP synthase (PEPS). Both can catalyze the conversion of pyruvate to PEP, although both reactions are reversible and thus also may catalyze the catabolic reaction from PEP to pyruvate. The difference lies in the used co-factors. PPDK requires P_i and produces AMP and PPi, while PEPS produces AMP and Pi to convert pyruvate and ATP to PEP. Interestingly, *C. acetobutylicum* does not contain a PPDK, while *C. beijerinckii* contains three isoenzymes, encoded by CBEI_0849, CBEI_3859, and CBEI_3862. Nevertheless, both species do contain three putative PEPS enzymes, encoded by CA_C0534,

CA_C0801, CA_C0797, CBEI_2063, CBEI_2066, and CBEI_3516. Currently, there are no expression or proteome data available that can identify the genes required for PEP synthesis by the PPDK (*C. beijerinckii*) or PEPS (*C. acetobutylicum* and *C. beijerinckii*)[35].

8.3.4 Pyruvate conversion

The oxidation of pyruvate can take three different routes, *viz.* via the pyruvate-ferredoxin-oxidoreductase (PFOR; EC 1.2.7.1), the pyruvate-formate-lyase (PFL; EC 2.3.1.54), and the pyruvate decarboxylase (PDC; EC 4.1.1.1). The oxygen sensitive PFOR is the major path for cleaving pyruvate to acetyl-CoA and CO_2 and transferring the electrons to ferredoxin[36]. *C. acetobutylicum* contains two *pfor* paralogs (CA_C2229, CA_C2499), both coding for large proteins (1171 and 1173 AA) and comprising the α, β, and γ domains. Again, no transcriptional or proteomic data are available that allow us to decide which is the most important enzyme for the formation of acetyl-CoA. Remarkably, *C. beijerinckii* has 3 *pfor* paralogs, also without experimental data to strengthen their role. Alternatively, pyruvate might be converted to acetyl-CoA by cleavage to formate as mediated by a pyruvate-formate-lyase. Among clostridia this enzyme has been detected in *C. kluyveri, C. pasteurianum*, and *C. butyricum*. Its role in clostridial metabolism is not clear and it has been suggested to be of anabolic rather than catabolic function[37,38]. *C. acetobutylicum* and *C. beijerinckii* also contain genes coding for this enzyme and it is interesting to note that *C. acetobutylicum* possesses only one gene (CA_C0980) while in the genome of *C. beijerinckii* two paralogs (CBEI_1009, CBEI_1011) are annotated as pyruvate-formate-lyase. These genes show high sequence similarity to the enzyme found in *C. pasteurianum*, for which it has already been shown to be active[37].

C. acetobutylicum is the only species among the solventogenic clostridia, that contains a gene coding for a pyruvate-decarboxylase, the enzyme that cleaves pyruvate into acetaldehyde and carbon dioxide. The gene (CA_P0025) is located on the pSOL1 megaplasmid. The physiologic role for this enzyme seems to be channeling pyruvate from glycolysis to ethanol without the requirement for CoA, specifically during the switch to solventogenesis when acids are taken up from the environment[39].

In addition to the different oxidative alternatives, pyruvate can be reduced to lactate involving lactate dehydrogenase (CA_C0267, CA_C3552, CBEI_1014, CBEI_4126 and CBEI_4972) or converted to acetolactate requiring acetolactate synthase (CA_C3169, CA_C3176, CA_C3652, CBEI_0212, CBEI_0217, CBEI_2646, CBEI_2686) and subsequently decarboxylated to acetoin involving acetolactate decarboxylase (CA_C2967). The latter gene has no ortholog in *C. beijerinckii*.

8.3.5 Acetyl-CoA conversion

Acetyl-CoA is a branch-point for the formation of acetate, the reduction to ethanol, and the condensation into carbon-4 compounds. Acetate formation occurs via the classical enzyme couple phosphotransacetylase (CA_C1742) and acetate kinase (CA_C1743), whose genes are clustered on the chromosome. For *C. beijerinckii* a similar situation exists, with both genes also clustered (CBEI_1164, CBEI_1165). However, next to this kinase, in both species, a small protein named acylphosphatase is encoded by CA_C2830 and CBEI_4233. This enzyme catalyzes the irreversible conversion of acetyl-P to acetate without formation of ATP. The same enzyme might also be active on the comparable substrate butyryl-P. Alternatively, it may also use 1,3 disphosphoglycerate as substrate and produce 3-phosphoglycerate. To date, no literature are available that indicate the functional expression of these enzymes in either of the solventogenic clostridia.

Alternatively, acetyl-CoA can act as a sink for reducing equivalents by the action of NAD(P)H-dependent alcohol-acetaldehyde dehydrogenases (AADH). For the reduction to acetaldehyde and ethanol, two paralogs exist in *C. acetobutylicum* and both solvent forming genes, designated as *adhE1* (or *aad*) and *adhE2*, are located on the megaplasmid (CA_P0162, CA_P0035, respectively). It has been shown that ADHE1 is active during solventogenic conditions, while ADHE2 is active during acidogenic conditions and becomes repressed during the shift to solventogenesis[39,40]. For *C. beijerinckii* only one *adhE* homolog (66%

identity) was recognized, located on the chromosome (CBEI_0305). In *C. acetobutylicum*, *adhE1* forms an operon (*sol* operon) with *ctfA* and *ctfB*, which code for the acetoacetyl CoA:acetate/butyrate:CoA transferase[41]. The *adhE* homolog in *C. beijerinckii* (CBEI_0305) is not clustered with *ctfA* and *ctfB* (CBEI_3833 and CBEI_3834, respectively). The latter two genes, however, form an operon with genes coding for the acetoacetate decarboxylase (*vide infra*) and an aldehyde dehydrogenase (CBEI_3832)[31,42]. The latter gene shows a low 31% identity with *adhE1* and *adhE2*. However, the fact that it is co-expressed with *ctfA/B* and *adc*, as in *C. acetobutylicum*, suggests that it represents the most likely *C. beijerinckii* counterpart of *adhE1*.

Two molecules of acetyl-CoA are converted to acetoacetyl-CoA by the action of acetyl-CoA acetyltransferase (thiolase, THL). In *C. acetobutylicum*, two THL isoenzymes exist, whose genes are located on the chromosome (*thlA*; CA_C2873) and on the megaplasmid (*thlB* or *thiL*, CA_P0078). Similar to *adhE2* and *adhE2*, *thlA* is stronger transcribed during acidogenic conditions, and *thlB* under solventogenic[39,43]. *C. beijerinckii* also harbors two thiolase genes (CBEI_3630, CBEI_0411*)*, with the former showing highest homology with *thlA*. Increased expression of a *thl* (CBEI_3630 or CBEI_0411) has been reported during the late acidogenic and solventogenic phase[44].

Acetoacetyl-CoA is reduced to 3-hydroxybutyryl-CoA by the action of 3-hydroxybutyryl-CoA dehydrogenase (HBD), which is encoded by CA_C2708 and CBEI_0325. In both species, this gene lies clustered with the genes for the subsequent steps, catalyzed by crotonase (CRT) and butyryl-CoA dehydrogenase (BCD), and identified as CA_C2712 and CA_C2711 in *C. acetobutylicum* and CBEI_0321 and CBEI_0322 in *C. beijerinckii.* Interestingly, in both species this gene cluster also includes two genes that code for two flavoproteins (EtfA, EtfB). Wang *et al.*[31] showed that both clusters (CA_C2708-2712, CBEI_0321-0325) do indeed form an operon. In *C. kluyveri* the EtfA/B flavoproteins are part of a protein complex, also comprising the BCD. This protein complex was shown to catalyze the NADH-dependent reduction of crotonyl-CoA to butyryl-CoA, and due to the exergonic nature of this reaction also enable the NADH-dependent reduction of ferredoxin[45-48]. Reduced ferredoxin can subsequently be used to produce H_2. The presence of such BCD/ETF clusters in *C. acetobutylicum* and *C. beijerinckii* also suggests that these bacteria are capable of crotonyl-CoA-dependent ferredoxin reduction using NADH. It has been demonstrated that the activity of BCD required the presence of EtfA/B[49] and that these components have to be expressed simultaneously[50]. *C. beijerinckii* appears to contain a second BCD/ETF cluster encoded by *bcd* (CBEI_2035), *etfA* (CBEI_2037), and *etfB* (CBEI_2036).

Butyryl-CoA can be used for the synthesis of butyrate, coupled to ATP synthesis, or it is reduced to butyraldehyde and butanol. Conversion to butyrate involves phosphotransbutyrylase (PTB) and butyrate kinase (BK), encoded by the clustered genes CA_C3076, CA_C3075, and CBEI_0203, CBEI_0204 for *C. acetobutylicum* and *C. beijerinckii*. In both species, paralogs exist for the butyrate kinase (CA_C1660, CBEI_4006, CBEI_4609), but for *C. acetobutylicum*, it has been discussed that this butyrate kinase might be involved in the reverse direction, i.e. the phosphorylation of butyrate[51]. Reduction of butyryl-CoA to butyraldehyde is catalyzed by the same set of alcohol dehydrogenases also reported for acetyl-CoA reduction (ADHE1, ADHE2). Reduction of butyraldehyde to butanol occurs using BDHI (CA_C3299) and BDHII (CA_C3298). *C. beijerinckii* contains only one BDH homolog, encoded by CBEI_2421.

As mentioned above, acetoacetyl-CoA can be converted to acetoacetate involving the CoA transferases *ctfA* (CA_P0163, CBEI_3833) and *ctfB* (CA_P0164, CBEI_3834). In addition, *C. beijerinckii* harbors a pair of *ctfA/B* paralogs (CBEI_2654, CBEI_2653). Acetoacetate is finally decarboxylated to acetone by an acetoacetate decarboxylase, encoded by *adc*. In *C. acetobutylicum*, this gene is part of the *sol* operon (CA_P0165). Similarly, its ortholog in *C. beijerinckii* (CBEI_3835) also lies clustered with *ctfA* and *ctfB* and a putative *adhE1*.

8.3.6 Energy generation and disposal of reducing equivalents

C. acetobutylicum and *C. beijerinckii* generate energy mainly by substrate level phosphorylation. They lack the ability to deposit electrons to terminal electron acceptors by membrane-bound electron transport chains. Therefore they need to regenerate their reduced electron carriers such as NADH or ferredoxins by producing H_2, or by forming reduced end-products such as lactate, ethanol, or butanol. However, the latter strategy has the disadvantage that the flux through acetyl-phosphate or butyryl-phosphate is diminished and accordingly less energy is conserved. Therefore, for each reduction equivalent deposited elsewhere, an additional substrate phosphorylation step can be done, optimizing energy yield. The solventogenic clostridia adopt different strategies for this optimization.

During sugar fermentation, reducing equivalents are produced at different levels. NADH is produced in the GAP dehydrogenase reaction (GAPDH) (Fig. 8.2). Reduced ferredoxin is produced during the oxidative decarboxylation of pyruvate (reaction (8.1)) and as discussed above, probably also during the exergonic reduction of crotonyl-CoA (reaction (8.2)).

$$\text{pyruvate} + \text{CoA} + 1\ \text{Fd}_{ox} \leftrightarrow \text{acetyl-CoA} + CO_2 + 1\ \text{Fd}_{red}^{2-} + 2\ H^+ \quad (8.1)$$

$$2\ \text{NADH} + 1\ \text{crotonyl-CoA} + 1\text{Fd}_{ox} \rightarrow 2\ \text{NAD}^+ + 1\ \text{butyryl-CoA} + 1\ \text{Fd}_{red}^{2} \quad (8.2)$$

These reduced electron carriers need to be recycled and hydrogenases play a central role in this. *C. acetobutylicum* as well as *C. beijerinckii* have been reported to produce H_2, with reported values of 2.0 mol H_2/mol glucose for *C. acetobutylicum*[52] and 1.6 mol H_2/mol xylose for *C. beijerinckii*[53] suggesting that ferredoxin-dependent hydrogenases should be present in both species. Such H_2 evolving hydrogenases are usually of the "iron-only" type ([Fe-Fe]) and these very oxygen-sensitive enzymes are predominant in Clostridia[54,55]. In contrast to other [Fe-Fe]-hydrogenases, the clostridial type enzymes contain three additional [4Fe-4S] and one [2Fe-2S] iron-sulfur clusters, which are involved in electron transfer to and from ferredoxin[56]. In *C. acetobutylicum*, *hydA* (CA_C0028) codes for the ferredoxin-dependent hydrogenase. *C. acetobutylicum* contains a second gene coding for another iron-only hydrogenase (*hydB*, CA_C3230). In comparison with HYDA, this enzyme shows a five-fold lower level of activity. Therefore, HYDA seems to be mainly responsible for H_2 formation[57]. Comparing the genomes of *C. acetobutylicum* and *C. beijerinckii*, it is interesting to note that *C. beijerinckii* contains no obvious ortholog for *hydA*, whereas one gene with low similarity to *hydB* (CBEI_1901) is present. *C. beijerinckii* contains two other putative hydrogenase genes, *viz*. CBEI_3796 and CBEI_4110. The former is clustered with 4Fe-4S-ferredoxin genes and the latter is part of a gene cluster (CBEI_4110-4112) encoding a trimeric bifurcating hydrogenase[58]. These recently described enzymes can form hydrogen from electrons coming from reduced ferredoxin and NADH together. Thereby, the strong reducing power of ferredoxin is combined with electrons coming from NADH regeneration to form hydrogen. These enzymes might have the disadvantage that hydrogen cannot be formed at the same partial pressure as with hydrogenases getting their electrons solely from ferredoxin, but might on the other hand help to regenerate some NADH.

C. acetobutylicum contains additional genes for a [Ni-Fe]-hydrogenase including a maturation protein (CA_P0141-CA_P0143), which are located on the pSOL1 megaplasmid.[14] Similar genes can also be found in *C. beijerinckii* (CBEI_3013-CBEI_3011). The physiologic functions of these [Ni-Fe]-hydrogenases are not yet known, but they can be speculated to be uptake hydrogenases because many [Ni-Fe]-hydrogenases have this function in other organisms. Previous studies have shown that hydrogenases are only active when they are co-expressed with hydrogenase maturation proteins which are responsible for the biosynthesis of the unique catalytic metallocluster[59-63]. In *C. acetobutylicum* genes coding for the maturation proteins *hypE* (CA_C0809), *hypF* (CA_C0810), and *hybG* (CA_C0808) have been recognized so far[57]. *C. beijerinckii* harbors a similar cluster (CBEI_3006 - 3009).

A major difference in the energy metabolism of *C. acetobutylicum* and *C. beijerinckii* is that *C. beijerinckii* possesses the genes coding for a membrane-bound Rnf complex (CBEI_2449-CBEI_2454)[64,65]. Such an Rnf complex catalyzes the ferredoxin-dependent reduction of NAD^+ and couples this to the build-up of a proton gradient over the cytoplasmic membrane. Remarkably, *C. acetobutylicum is* one of the few sequenced clostridia without such a complex. Therefore, *C. beijerinckii* seems to be able to conserve energy in the form of a proton motive force by reoxidation of reduced ferredoxin whereas in *C. acetobutylicum* the reoxidation of ferredoxin has to be mediated mainly by the H_2 forming hydrogenase HydA (CA_C0028). Hydrogen might be formed by HydA at a higher partial pressure than with the bifurcating hydrogenase of *C. beijerinckii*. A soluble ferredoxin: NAD(P)H-oxidoreductase activity has been postulated in the cell extract of *C. acetobutylicum*, but the corresponding gene has not yet been identified and a physiologic role apart from probably biosynthetic NADPH formation of this activity is obscure, as it would result in energy dissipation connected to unfavorable NADH accumulation. The reason for the mentioned differences is not yet clear although one can speculate that *C. acetobutylicum* might be able to form more H_2 from ferredoxin which might allow for larger amounts of ATP formed via substrate level phosphorylation.

At this point of the discussion it should be clear that electrons in the form of reduced ferredoxin are of special value for a fermentative organism because it is possible to dissipate a large part of them as H_2 or to channel them through an Rnf complex to translocate protons. If H_2 is formed, more energy can be conserved during fermentation, because a larger amount of acid-phosphates can be converted to acids with the generation of ATP, this comes at the price of more pronounced acidification of the medium. If an Rnf complex is present, electrons from reduced ferredoxin can also be used for ATP generation via the proton motive force and the ATPase. But this comes with the price of the NADH formed during Rnf reaction, which in turn has to be regenerated. One might also speculate that the membrane-bound Rnf complex could easily become a rate limiting step when compared to the soluble acid phosphate kinases.

These facts might be important for proposing the differences in energy metabolism between *C. acetobutylicum* and *C. beijerinckii*. *C. acetobutylicum* optimizes the rate of energy production by dissipating much of its electrons from reduced ferredoxin as molecular H_2. Because less NADH has to be regenerated, this allows fast growth due to a high rate of ATP generation by soluble enzymes, but results in rapid acetate and butyrate formation and acidification of the medium. The formation of solvents is a second step when the formed acids are partly taken up again in order to increase pH to buy time for sporulation. Accordingly, the solvent formation genes are located on the megaplasmid pSOL1, as solvent formation for *C. acetobutylicum* represents mainly a metabolic "extension" for habitats where rapid acidification poses a problem for sporulation.

C. beijerinckii in turn might channel part of its reduced ferredoxin to the Rnf complex. This directly generates energy and allows a slower rate of acid production. A consequence of this mode of energy production is that more NADH needs to be regenerated as less ferredoxin is directly used for H_2 production and the ferredoxin conversion at the Rnf complex results in NADH generation. Therefore more reduced end-products like ethanol, butanol, or isopropanol should to be formed when the Rnf complex is used. Interestingly some strains of *C. beijerinckii*, but not *C. acetobutylicum*, contain an isopropanol dehydrogenase activity for reduction of acetone allowing additional NADH regeneration[7,66]. Additional NADH could be regenerated by H_2 formation using the trimeric bifurcating hydrogenase[58]. But this method of H_2 formation is energetically less efficient than H_2 formation solely from ferredoxin. Solvent formation seems to be much more integrated in normal growth mode and accordingly genes for solvent formation are located directly on the chromosome in *C. beijerinckii*.

8.4 Conclusion

As the overall acidogenic and solventogenic metabolisms of *C. acetobutylicum* and *C. beijerinckii* are very similar, one can expect that similar pathways and enzymes are involved. Indeed, for most of the enzymes that participate in the central glycolytic pathway of *C. acetobutylicum*, at least one ortholog in *C. beijerinckii* exists. Moreover, genes that are clustered in *C. acetobutylicum* often also lie clustered in *C. beijerinckii*, e.g. the glycolytic carbon-3 cluster, the BDH/ETF cluster, and the kinase/PTA/PTB cluster. Essential differences appear to exist with respect to the pyruvate decarboxylase (only in *C. acetobutylicum*) and the Rnf complex as well as the trimeric bifurcating hydrogenase (only in *C. beijerinckii*). The physiological meaning of these differences is, however, not yet fully understood.

Another major difference is the presence of the megaplasmid in *C. acetobutylicum*, carrying important genes for solvent production. *C. acetobutylicum* appears to be better equiped for solvent production, as it has four ADHs (i.e. AAD, ADHE2, BDHI, and BDHII), whereas C. beijerinckii has only two orthologs (i.e. CBEI_0305 and CBEI_2421). In contrast to the latter example, in general *C. beijerinckii* often contains more (clustered) paralogs (especially alcohol dehydrogenases) than *C. acetobutylicum*. The reason for this is not clear. Moreover, the identity of the genes responsible of the major fermentative steps in *C. beijerinckii* is mainly based on homology to *C. acetobutylicum* genes and gene clusters. To prove the identity of the key genes, genome-wide microarrays and/or proteomic data are required.

Acknowledgement

This work was supported by the Netherlands Organization of Scientific Research (NWO; 826.06.003 and 826.09.001) and the German BMBF (Forschungvorhaben 0315782) through the SysMO project COSMIC (http://www.sysmo.net/).

References

1. Jones D. T. and Woods D. R., *Microbiol. Rev.* 50 (1986).
2. Dürre P., *Appl. Microbiol. Biotechnol.* 49 (1998).
3. Mitchell W. J., *Adv. Microb. Physiol.* 39 (1998).
4. Keis S. *et al., Int. J. Syst. Evol. Microbiol.* 51 (2001).
5. Dürre P., *Biotechnol. J.* 2 (2007).
6. Gheshlaghi R. *et al., Biotechnol. Adv.* 27 (2009).
7. George H. A. *et al., Appl. Environ. Microbiol.* 45 (1983).
8. Johnson J. L. and Chen J. S., *FEMS Microbiol. Rev.* 17 (1995).
9. Bahl H. and Gottschalk G., *Biotechnol. Bioeng. Symp.* 14 (1984).
10. Wilkinson S. R. *et al., FEMS Microbiol. Rev.* 17 (1995).
11. Gottschal J. C. and Morris J. G., *Biotechnol. Lett.* 3 (1981).
12. Woolley R. C. and Morris J. G., *J. Appl. Bacteriol.* 69 (1990).
13. Cornillot E. *et al., J. Bacteriol.* 179 (1997).
14. Nölling J. *et al., J. Bacteriol.* 183 (2001).
15. Keis S *et al., Int. J. Syst. Bacteriol.* 45 (1995).
16. Thompson J. D. *et al., NucleicAcids Res.* 25 (1997).
17. Young M. *et al., FEMS Microbiol. Rev.* 63 (1989).
18. Mermelstein L. D. *et al., Biotechnology* 10190 (1992).

19. Brouns S. J. *et al.*, *Science* 321 (2008).
20. Gevers D. *et al.*, *Trends Microbiol.* 12 (2004).
21. Mushegian A. R. and Koonin E. V., *Trends Genet.* 12 (1996).
22. Tatusov R. L. *et al.*, *BMC Bioinformatics* 4 (2003).
23. Rafil F. *et al.*, *Appl. Environ. Microbiol.* 57 (1991).
24. Liyanage H. *et al.*, *J. Mol. Microbiol. Biotechnol.* 2 (2000).
25. Mitchell W. J. and Tangney M., in *Handbook on Clostridia*, Dürre P., Ed. (CRC Press, Boca Raton, 2005).
26. Tangney M. and Mitchell W. J., *Appl. Microbiol. Biotechnol.* 74 (2007).
27. Tomas C. A. *et al.*, *J. Bacteriol.* 185 (2003).
28. Jones S. W. *et al.*, *Genome Biol.* 9, R114 (2008).
29. Sullivan L. and Bennett G., *J. Ind. Microbiol. Biotechnol.* 33 (2006).
30. Schreiber W. and Dürre P., *Anaerobe* 6 (2000).
31. Wang Y. *et al.*, *BMC Genomics* 12 (2011).
32. Iddar A. *et al.*, *Prot. Expr. Purif.* 25 (2002).
33. Hillmann F. *et al.*, *FEBS Lett.* 583 (2009).
34. Alsaker K. V. *et al.*, *J. Bacteriol.* 186 (2004).
35. Mao S. *et al.*, *J. Proteome Res.* 9 (2010).
36. Meinecke B. *et al.*, *Arch. Microbiol.* 152 (1989).
37. Weidner G. and Sawers G., *J. Bacteriol.* 178 (1996).
38. Wood N. P. and Jungermann K, *FEBS Lett.* 27 (1972).
39. Grimmler C. *et al.*, *J. Mol. Microbiol. Biotechnol.* 20 (2011).
40. Fontaine L. *et al.*, *J. Bacteriol.* 184 (2002).
41. Fischer R. J. *et al.*, *J. Bacteriol.* 175 (1993).
42. Chen C. K. and Blaschek H. P., *Appl. Environ. Microbiol.* 65 (1999).
43. Winzer K. *et al.*, *J. Mol. Microbiol. Biotechnol.* 2 (2000).
44. Shi Z. and Blaschek H. P., *Appl. Environ. Microbiol.* 74 (2008).
45. Herrmann G. *et al.*, *J. Bacteriol.* 190 (2008).
46. Li F. *et al.*, *J. Bacteriol.* 190 (2008).
47. Seedorf H. *et al.*, *P. Natl. Acad. Sci. USA* 105 (2008).
48. Köpke M *et al.*, *P. Natl. Acad. Sci. USA* 107 (2010).
49. Boynton Z. L. *et al.*, *J. Bacteriol.* 178 (1996).
50. Inui M *et al.*, *Appl. Microbiol. Biotechnol.* 77 (2008).
51. Huang K. X. *et al.*, *J. Mol. Microbiol. Biotechnol.* 2 (2000).
52. Warner J. B. and Lolkema J. S., *Microbiol. Mol. Biol. Rev.* 67 (2003).
53. Ye X. *et al.*, *Appl. Microbiol. Biotechnol.* 92 (2011).
54. Erbes D. L. *et al.*, *Plant Physiol.* 63 (1979).
55. Vignais P. M. *et al.*, *FEMS Microbiol. Rev.* 25 (2001).
56. Peters J. W. *et al.*, *Science* 282 (1998).
57. King P. W. *et al.*, *J. Bacteriol.* 188 (2006).
58. Schut G. J. and Adams M. W. W., *J. Bacteriol.* 191 (2009).
59. Blokesch M. *et al.*, *J. Mol. Biol.* 344 (2004).
60. Blokesch M. *et al.*, *Biochem. Soc. Trans.* 30 (2002).
61. Drapal N. and Bock A., *Biochemistry* 37 (1998).
62. Lutz S. *et al.*, *Mol. Microbiol.* 5 (1991).

63. Posewitz M. C. *et al.*, *J. Biol. Chem.* 279 (2004).
64. Biegel E. and Muller V., *P. Natl. Acad. Sci. USA* 107 (2010).
65. Biegel E. *et al.*, *Cell. Mol. Life Sci.* 68 (2011).
66. Hiu S. F. *et al.*, *Appl. Environ. Microbiol.* 53 (1987).
67. Markowitz M. *et al.*, *Nucleic Acids Res.* 34 (2006).
68. Alsaker K. V. and Papoutsakis E. T., *J. Bacteriol.* 187 (2005).
69. Tomas C. A. *et al.*, *J. Bacteriol.* 186 (2004).
70. Tummala, S S. B. *et al.*, *Biotechnol. Bioeng.* 84 (2003).
71. Paredes C. J. *et al.*, *Nucleic Acids Res.* 32 (2004).
72. Tomas C. A. *et al.*, *Appl. Environ. Microbiol.* 69 (2003).
73. Fiedler T. *et al.*, *J. Bacteriol.* 190 (2008).
74. Zhao Y. *et al.*, *Appl. Environ. Microbiol.* 71 (2005).
75. Kosaka T. *et al.*, *J. Bacteriol.* 188 (2006).
76. Tummala S. B. *et al.*, *J. Bacteriol.* 185 (2003).
77. Gorwa M. F. *et al.*, *J. Bacteriol.* 178 (1996).

CHAPTER 9

THE STRATEGIC IMPORTANCE OF BUTANOL FOR JAPAN DURING WWII: A CASE STUDY OF THE BUTANOL FERMENTATION PROCESS IN TAIWAN AND JAPAN

D. T. JONES

Department of Microbiology and Immunology
University of Otago Dunedin, New Zealand

During the first half of the 20th century the industrial acetone-butanol-ethanol (ABE) fermentation was second only in importance to the industrial ethanol fermentation. The strategic importance of the fermentation process for the United Kingdom, France and the United States during World War I, and the subsequent expansion of commercial fermentation in a number of countries in the Western world, has been extensively documented. More recently information regarding the industrial processes that operated in Russia and China has also been published. The butanol fermentation is also known to have been operated in Taiwan and Japan before, during and after World War II. It is apparent that the large-scale industrial fermentation process that was established in Taiwan and Japan was of significant strategic importance during the war. Commercial butanol fermentation processes were also in operation in both Japan and Taiwan after the war. However, there is very little information available regarding what was undoubtedly a major industrial fermentation industry in the Far East. This review attempts to draw together disparate information from a variety of sources to provide an overview of this largely overlooked facet of the industrial butanol fermentation process. The strategic importance of this fermentation during World War II, which formed the basis of the subsequent establishment of commercial processes, provides an interesting case study with some relevance for the current renewed interest in the production of biobutanol for use as biofuel and chemical feedstock.

9.1 Introduction

The resurgence of interest in biobutanol as a chemical feedstock and biofuel has led to a re-evaluation of the industrial ABE fermentation process. The fermentation was of strategic importance during World War I (WWI), for the United Kingdom (UK), France and the United States of America (US). The allies depended on acetone as a solvent for manufacturing munitions and, to a lesser extent, for aircraft dope. The key role played by Weizmann in the development of the industrial ABE fermentation process in the UK, and its subsequent transfer to Canada and the US has been well documented[1,2].

The patented commercial process that was first operated in the US after WWI, and its subsequent expansion globally, has also been extensively reviewed[2,3,4,5]. Technical information is available on the commercial operation of the process in the US[6,7,8], the UK[9,10] and South Africa[11,12]. More recently, information regarding the processes that operated in Russia[13] and China[14] have appeared in the literature. The ABE process was also operated on a smaller scale in Brazil and Egypt, but no technical details regarding these processes have been published. Industrial butanol fermentation is also known to have been in use in Taiwan and Japan before, during and after World War II (WWII). However, there is very little information available in the scientific literature with respect to the nature and operation of what was apparently a large-scale industrial fermentation. The meagre information that is available provides a tantalising indication that this was of significant strategic importance for Japan. The expansion of the petrochemical industry after WWII led to a worldwide decline in the uses of the industrial fermentation process during the second half of the 20th century and the process had become obsolete well before the end of the century.

The scarcity of information in the scientific literature regarding the butanol fermentation process that was developed by Japan and operated in Taiwan and Japan prior to, during and after WWII, required the sourcing of information from a wide range of alternative sources. Much of this supplementary information has been retrieved from websites providing access to reports, book chapters and military archives, along with material gleaned from various websites and blog discussions. Information obtained from these sources can often be difficult to verify. However, by piecing together disparate material from a wide range of sources, it has been possible to assemble a comprehensive overview of the strategic importance of butanol fermentation for Japan during WWII, as well as its post-war commercialisation in both Taiwan and Japan. The background information contained in the first three sections has been condensed from a wide range of sources. This information is widely known, and is readily accessible, so specific references for this introductory material have largely been omitted. Almost all of the information contained in the section dealing with the acquisition of US petroleum technology by Japan was abstracted from a single source[15]. Much of the information underpinning this account was sourced from WWII and post-war US military documents, intelligence reports and other material dealing with Japanese and Taiwanese wartime fuel supply, the identification and location of butanol plants, bombing targets and the effects of the bombing campaign. This was used to piece together an overview of the strategic importance of butanol production to Japan during WWII. Many helpful authored references were written during the post-war period[16,17,18,19]. Previously classified reports (now declassified) prepared for the US Army and Navy, either during the war or as part of the post-war assessments and analysis proved invaluable. This information was supplemented with information retrieved from a number of websites devoted to air force and aviation fuel history.

The majority of websites do not specify the author or the date. The most extensive and useful reference material was retrieved from a number of websites[20,21,22,23,24,25]. Numerous other retrievals from websites containing either brief or relatively minor items of information have not been individually referenced. Another valuable reference source, originally prepared for the US Quartermaster General's Office, is a report on WWII Formosa Japan Oil Fields[26]. An opportunity to visit the Refining & Manufacturing Research Institute of the CPC Corporation in Taiwan, which occupies the site of the original butanol plant, provided a great deal of unpublished information and photographic material. In-house documents, providing a history of the Godo Shusei Company in Japan were generously provided by Oenon Holdings, Inc. This provided a large amount of additional information regarding the history of the ABE fermentation process in Japan.

In view of the recent interest in biofuels and biobutanol by the US military, airlines and the oil and chemical industry, the role of biobutanol in the production of aviation fuel during WWII provides an informative case study. The historical evidence points to Japan's limited access to crude oil and aviation fuel as one of the major factors leading to her defeat in WWII. If another global conflict were to break out in the future, access to adequate liquid fuel supplies for military purposes could again become a key factor in determining the outcome.

9.2 Historical events that led to the establishment and development of the ABE fermentation process in Japan prior to WWII

Following the restoration of the Meiji Empire in 1868, Japan embarked on major changes to what had previously been an isolated, inward-looking, feudal empire. Although as a country, Japan lacks many essential natural resources, including petroleum, coal, iron ore, aluminium, copper, zinc and rubber, it underwent a rapid programme of modernisation, and transformed itself from a virtual medieval state to a modern industrial power in the space of only four decades. The development of its industrial capabilities coincided with the expansion of the Japanese empire. Victory over China in 1895 resulted in Japan acquiring the island of Taiwan (Formosa). The defeat of the much stronger Imperial Russian fleet in 1905 opened the way for Japan to annex Korea in 1910. During WWI Japan became an ally of the UK and the US and, following the peace conference in 1919, it was given a mandate over a large area of the Pacific. Japan's rapid industrial growth continued, and by 1927 its trade in China had surpassed that of the UK. As Japan's industrial base expanded it became increasingly dependent on the importation of essential raw materials. At this time concerns regarding US naval expansion provided the stimulus for Japan to embark on an extensive naval building programme to safeguard its expanding empire.

The Great Depression, at the beginning of the 1930s, triggered a major setback for Japan, leading to massive layoffs at industrial plants and high rate of unemployment. Crop failures also resulted in famine. The economic crisis precipitated a revolt by war-hungry army officers, supported by a strident navy faction. As a result, extreme nationalist factions gained power, culminating in the increasing militarisation of the Japanese Empire. The impact of the depression further emphasised

Japan's vulnerability with respect to her reliance on indispensable imports and this became a major driving force for Japan's militaristic government embarking on a mission of territorial expansion. Japan maintained a garrison in Manchuria as a defensive measure against possible invasion by Russia, and in 1931 the Army staged an incident that enabled them to justify taking over the coal and iron-rich territory. The subsequent signing of a truce with China ceded the control of Manchuria to Japan and the state of Manchukuo was established in 1932. By 1935 Japan's trade had reached record levels and the country had begun converting to wartime production. In 1936, Japan withdrew from the London Naval Conference, and embarked on a renewed naval arms race. By this time Japan had come to the view that most of Asia should rightfully be incorporated within the Japanese Empire. The cabinet endorsed three guiding principles: the first was to drive Western powers out of Asia, the second was to develop more extensive interactions with other Asian powers and the third was to create an economic bloc that encompassed China. However, Japan was thwarted in its goal of unifying and dominating Asia due to its lack of natural resources, as well as the continued presence of Western powers in the region. This included the UK in Malaysia, Burma and Hong Kong, the Dutch in the East Indies and the US in the Philippines. To further its objectives of Asian expansion, Japan embarked on the invasion of China in 1937.

9.3 The key role played by Japanese air power

Following the Meiji restoration, the Japanese military began diligently and methodically following military and technical developments in other countries. Early on Japan began investigating the use of aircraft as a potential weapon, and during WWI they acquired examples of several wartime aircraft types. After WWI, a delegation from the British Royal Navy recommended the establishment of an Imperial Japanese Navy Air Force, and helped with its founding and officer training. A second delegation from Germany assisted with the establishment and training of an Imperial Japanese Army Air Force. Up until the early 1930s, the two Japanese air services were largely equipped with obsolete foreign aircraft types, either imported or built under license in Japan. Around this time,

Japanese aircraft designers began producing home-designed aircraft types that were better adapted to their own operational requirements. Each air service had different imperatives: the Army was preparing for war against China and the Soviet Union, while the Navy was preparing itself for warfare against the US and UK in the western Pacific. As a consequence, each service developed its own air arm tailored to meet its particular needs. To the overall detriment of the later Japanese war effort, neither service cooperated with the other. Army and Navy aircraft employed different technical systems, and aircraft factories were divided between those that built aircraft for the Army and those producing naval aircraft. There was little exchange of information, and each service kept its designs and technical developments secret from the other.

At Admiral Yamamoto's insistence the Navy's air service was built up to a formidable strength. By the time war was declared in December 1941, Japan had 11 aircraft carriers and the Navy air arm possessed in excess of 2,000 aircraft, of which around a quarter were carrier-based with the fleet. These consisted of a number of excellent aircraft types that included the highly manoeuvrable A6M Zero fighter as well as torpedo bombers and dive bombers. The Navy also possessed the Long Lance torpedo that was the most advanced in the world. At this time the Japanese naval air arm was the most highly developed in the world, and Japanese pilots were among the best trained. However, due to Japan's isolation from the West, and the general secretiveness of the government and society, these important developments were not recognised or appreciated. The Japanese air forces met with considerable success in the Sino–Japanese War, where they dominated the skies over China and instituted the strategic bombing of Chinese cities. Even so, there was a widespread ignorance and disinformation in the West resulting in an expectation that, in the event of war, the allied air forces would encounter a small force of obsolete Japanese aircraft, flown by inferior, poorly trained pilots. As a result the allies totally underestimated Japan's capability, and Japanese air power came to dominate during the early stages of the war in the Pacific. Control of the air proved to be a vitally important aspect of naval operations in WWII. However, neither side fully appreciated how vulnerable battle ships would be to attack by aircraft.

9.4 The importance of high-octane aviation fuel

After WWI, air force strategists came to the conclusion that the limiting factors in future aerial warfare would not only depend on the ability to build aircraft and train pilots, but would be largely dependent on the capacity to produce aircraft fuel. It is therefore not surprising that the development and supply of high performance aviation fuel emerged to be of at least equal importance to other areas of aircraft design and engine development between the wars.

Petroleum provides the base feedstock for all aviation fuels. Crude oil comes in various grades, and consists of mixtures of numerous straight and branched chained hydrocarbons of varying molecular weights, associated with assorted impurities, notably sulphur-containing compounds. Initially distillation provided a straightforward way of separating the fractions in crude oil and only required relatively simple technology. However, distillation can only separate out from the crude whatever fractions happen to be present, in more or less their existing proportions, rather than the fractions that the refiner desires the most. Before the advent of the internal combustion engine, the fraction of greatest value was kerosene, widely used in lamps, which comprised around 60% of crude oil. Gasoline, consisting a complex mixture of hydrocarbons containing between C5 and C10 carbon atoms, was too volatile for use in lamps, and only had a small market. This changed rapidly with the advent of the automobile. By 1910 there were already around 500,000 motor vehicles on the road, and the demand for gasoline skyrocketed. As a consequence, refiners were desperate to find ways to obtain more gasoline from a barrel of crude oil.

For the first few decades of flight, aircraft engines utilised the same type of gasoline that powered automobiles. During WWI, pilots noticed that gasoline refined from Romanian crude oil performed better than that refined from California crude. It was subsequently found that the straight chain alignment of carbon atoms in gasoline distilled from crude oil combusts very easily. As a result, the compression of a gasoline/air mix causes detonation to occur before the piston reached the top of its stroke. This produces two explosions per stroke, one by compression and one by spark ignition. This double explosion, known as knocking, greatly interferes with the efficient working of the engine. The more powerful the engine, the greater the compression and thus the greater the problem with premature ignition. In 1919, Kettering and Midgley investigated the source of the knocking, and experimented with different compounds to try and prevent it. After extensive testing they discovered that tetraethyl lead (TEL) could be used as an effective anti-knock agent that could also be produced at an economical price. The major disadvantage was that this compound proved to have very harmful environmental effects. In 1926, Graham discovered that knocking could be suppressed by adding iso-octane to gasoline. He developed the octane rating scale, using heptane and iso-octane as reference fuels that were given the arbitrary octane numbers of 0 and 100, respectively. Initially the gasoline being produced had an octane rating of between 25 and 50 but by the late 1920s the addition of TEL led to significant improvements. At this time aviation gasoline had an octane rating of between 75 and 80. Aviation gasoline with such high octane numbers could only be produced by the distillation of high-grade petroleum. However, even this quality gasoline was no longer adequate as fuel for the larger, more powerful engines used by piston-driven aircraft that were developed in the 1930s and 1940s.

Attempts to increase the yield of gasoline from each barrel of oil led to the development of 'thermal cracking'. In this process, crude oil was heated at high pressure and temperature to 'crack' long-chain molecules into smaller ones. The reformed molecules form branched chains that are more resistant to knocking, so that both the yield and the octane rating was improved. Standard Oil of Indiana first began investigating thermal cracking around 1909, and in 1913 the company patented the Burton–Humphrey batch process. This was used in the company's refineries to double gasoline yields to 25%. Other refiners either licensed the process from Standard of Indiana or came up with their own versions of cracking. The thermal cracking process then remained the state of the art technology for the next 20 years. By the end of the 1920s, a series of advances, which included more efficient continuous flow processes, resulted in typical gasoline yields exceeding 40%. The quality of the gasoline was also improved, though not as much as the quantity, but with the addition of TEL it was adequate for most motor vehicles.

Increasing the octane number of gasoline became a major concern as the demand for more powerful engines grew. Continuing improvements in aircraft design after WWI led the US Army to specify fuel of 100 octane or higher for its fighter planes. At the time most aviation gasoline had no more than an 87 octane rating. In the 1930s, Major James Doolittle was appointed to head the aviation fuels section of the Shell Oil Company. He was already famous in the aviation community as a racing pilot, and for his support of advanced research and development. He realised that if the US were to become involved in another major war, it would require large amounts of high octane aviation fuel. Doolittle pushed hard for the development of high octane aviation gasoline, and, with his encouragement, Shell embarked on the technical development of 100-octane fuel. They found hat the combined use of iso-octane and TEL as fuel additives made it possible to produce 100-octane fuel. However, this new fuel remained a scientific curiosity, as there was no feasible means for a refinery to economically produce it in bulk quantities.

A breakthrough came with the development of the alkylation process for the production of iso-octane by the British-owned Anglo-Persian Oil Company in 1935. In this process iso-butylene is either first dimerised, or alkylated to iso-butane, using a strong acid catalyst. This yields a mixture of iso-octenes that are then hydrogenated to produce iso-octane. In 1938 Humble Oil in the US paid over $8,000,000 to licence the new alkylation process, which allowed iso-octane to be produced on a large scale for the petroleum industry. This allowed the blending of refined gasoline, with an octane number in the 70s, with pure iso-octane and iso-pentane to boost that octane number. This was then boosted still further using additives such as TEL. The addition of TEL was highly effective, but there was a limit to how much of it could be added without leaving deposits on the pistons and cylinders. Doolittle convinced Shell to begin manufacturing the fuel, and to stockpile the necessary chemicals and modify its refineries to make mass production of high-octane fuel possible. As a result, when the US entered the war in late 1941, it had large quantities of high quality fuel for its air forces. At the beginning of WWII, Shell manufactured 25% of the 100-octane aviation gasoline used by the military. By 1943 a cold alkylation process had been developed that produced enormous outputs of the 100-octane AvGas.

A further major advance came about through the development of a catalytic cracking process[27]. Catalytic cracking had first been investigated in the 1920s by Standard Oil of New Jersey, but research on it was abandoned during the Depression. A French engineer named Houdry, financed by the Sun Oil Company, undertook trials at their Pennsylvania plant which demonstrated that gasoline produced by Houdry's catalytic process had an octane number of 81, which was around 8 better than the best additive-enhanced gasoline at this time. From 1933 to 1935, the engineering team set about solving the many technical challenges, and his new catalytic refining method was patented in 1937. The newly built plant was capable of refining 15,000 barrels a day with octane numbers between 77 and 81. The process did not require high-grade crude but used the thick oil residues that were normally converted to fuel oil. At the time most refineries were producing 60 octane boosted to 72 by adding TEL. The new process increased gasoline yield from 25 to 50% and among its virtues, catalytic cracking produced gasoline with a lower sulphur content than other methods, making it more responsive to TEL's effects. By starting with this reformed gasoline, aviation fuel could be enhanced to 100 octane using only half the additives needed for other types of gasoline. By 1941 Houdry's firm had developed a new catalyst that could produce aviation-quality gasoline with one pass through the catalytic unit, eliminating the original two-step reforming process. Houdry's process produced 90% of all catalytically cracked allied aviation fuel in the first two years of US involvement in WWII. From a start of 40,000 barrels per day in 1941, production climbed to 200,000 barrels per day in 1943 and peaked at 373,000 barrels per day in 1944. In the late 1930s and early 1940s, when the success of Houdry's process became apparent, Standard of New Jersey resumed the project as part of a consortium with Standard of Indiana, Universal Oil Products, Shell and Texaco. The consortium developed a continuous process, using a catalyst ground into a fine powder and suspended in a flow of air in a fluidised bed that had vapours of the petroleum feedstock blown through it. Without catalytic cracking to produce a base stock of high-octane gasoline to mix with the additives and blending agents it would not have been possible to meet aviation gasoline requirements for the allied air forces. Although the fluid catalytic-cracking processes made the original Houdry method obsolete, the original Houdry process had both established the use of catalysis in the petroleum industry and helped win the war for the allies[27].

9.5 Japan's vulnerability with respect to oil and aviation fuel

The Japanese were all too aware of their vulnerability with regard to oil and amongst their greatest concerns was their dependency on the US for the importation of crude oil, petroleum and, in particular, high-octane aviation fuel. In an attempt to achieve independence in liquid fuel supply, Japan embarked on a dual strategy. One major objective was to gain access to the oil fields in Southeast Asia.

The second major objective was to establish a synthetic fuel industry in Japan and her colonies. Japanese research into synthetic fuel production from coal was first initiated in the 1920s, only a few years after other countries, such as Germany and Britain, embarked on the quest for the conversion of coal to oil. Japanese technological developments for the conversion of coal to oil were based mainly in coal rich Manchuria, and extensive laboratory research was undertaken on the coal hydrogenation and Fischer–Tropsch conversion processes[28]. However, in their haste to construct large synthetic fuel plants, the key intermediate pilot-plant stages were bypassed. As a result, the transition from small-scale to large-scale production was never successfully achieved. Although Japan was unable to develop a programme to synthesise liquid fuels from coal, they were able to develop processes to produce significant quantities of liquid fuels from the technologically simpler coal carbonisation and oil-shale distillation processes[19].

During the 1930s Japan's Navy, along with the Japanese Government, acquired a German manufacturing license for the installation for a shale plant at Fushun, in Manchukuo, capable of an annual production of around 200,000 tons of shale oil. In 1937 Japan embarked on a seven-year plan that called for completion of 66 carbonisation plants using coal as a feedstock, 10 hydrogenation plants using coal tar and shale oil distillates and 11 plants using the German Fischer–Tropsch hydrocarbon synthesis process. It was hoped that these synthetic facilities would produce 6,400,000 barrels of gasoline and 7,700,000 barrels of heavy oil products annually[19]. By the time war broke out only eight plants were in action. Production of synthetic oil from coal was a major disappointment for Japan, and only reached a maximum rate of some 114,000 tons per annum. The oil shale works at Fushun in Manchuria produced more oil than the combined production from all the coal to oil plants in the Japanese Empire.

Seeking a solution to the problem of Japan's over-dependence on imported aviation fuel, the Japanese Navy initiated studies on the production of alcohol and other synthetic chemicals from renewable resources. In the early 1920s the Navy commissioned research on techniques for producing alcohols by fermentation for use as fuel. Extensive tests were carried out by the Navy on the use of ethanol, which allowed engines to run cooler as well as at higher compressions, and also permitted a higher rpm to be achieved. However, a number of difficulties and drawbacks were encountered regarding the use of ethanol as aviation fuel. These included low volatility, low anti-knock properties and its corrosive action on metals. The Navy also experimented with the use of methanol as aviation fuel, to take advantage of its extremely high latent heat of vaporisation. Later work was undertaken on the feasibility of producing methanol from CO_2 and H_2, which were produced as waste gases from the ABE fermentation process. However, the process was never brought into use because the the ABE fermentation programme was terminated before this could be implemented. The Japanese also tested various other substitutes for use as aviation fuel including ethyl ether, isopropyl ether, acetone, alkyl-benzenes, pine root oil and camphor oil. With the exception of a small amount of fuel refined from pine root oil used by the Navy at the close of the war, none of these were ever used as fuels.

9.6 The establishment of the Japanese ABE fermentation industry

Around 1920, the Japanese Navy initiated a programme to investigate the production of butanol by fermentation, as another potential component of aviation fuel. This work was entrusted to the Agricultural Department of Tokyo Imperial University. Bacterial cultures, including the Weizmann BY strain, were imported from the UK and cultures of solvent-producing

Clostridium were also isolated and selected locally, including the Rokushu No. 1 strain. Examples of the Rokushu strain dating from 1922 and the BY and BGP strains dating from 1924 and 1930 were later investigated by Rokushu & Yamazaki[29].

Hokkaido, the northern-most island of Japan, has nearly one-fourth of Japan's total arable land, and ranked first in the nation in the production of a range of agricultural products that included wheat, soybeans, potatoes and sugar beet. In 1927, investigations into the uses of industrial ABE fermentation process was initiated at two research institutes on Hokkaido (the Hokkaido and Sapporo Kogyo Shikenjo). Sweet potatoes were identified as the main fermentation substrate for the process. A number of culture collection strains were tested and the most promising strains were selected for trials. The culture best fitted for the purpose proved to be the number 314 strain. This strain was used to investigate the conditions for optimizing the ABE fermentation, using sweet potatoes. A maximum performance of 21.65 g/l total solvent with 15.6 g/l butanol and 9.6 g/l acetone at a yield of 31.2% was reported to have been achieved from a total sugar concentration of around 5%. Optimisation for sweet potatoes was achieved by regulation of the pH and the addition of an appropriate nitrogen source such as soy bean cake. Pilot-scale trials were undertaken, but these were stopped in 1930. Although the trials had produced favourable results, the process was not commercialised at this time. Investigations into the production of acetone and butanol by fermentation continued at the Institute of Industrial Technology in Kotoni, located near Sapporo, using a Japanese isolate named *Bacillus butyloacetonicus* (Japanese patent No. 101,128, 1933). Corn was found to be the most suitable substrate for this strain. In 1928, studies on industrial ABE fermentation were also begun in the Agricultural Department of Kyushu University.

In 1931, work was initiated on the building of an industrial-scale ABE fermentation plant by the Kyoei Company, with assistance provided by the Saki-shi Kondo Pharmaceutical Company, in the Osaka area. The initial production of butanol at this plant was 20 tons per month, but by 1933 production had been doubled to 40 tons a month. By 1935, Japan was producing 1,000 tons of butanol a year, and this was increased to 2,000 tons in 1937.

In June 1936, three Japanese alcohol distillers (Takara Shuzo, Godo Shusei and Dainippon Shurui), amalgamated to form a consortium that later became the Kyowa Hakko Kogyo company. Members of the consortium all specialised in the production of commercially marketed fermented products, mainly ethanol, that was used in sake and other alcoholic beverages. In November 1937 the relationship between the three companies was formalised by founding the Kyowa Chemical Research Laboratory in Tokyo. The government was aware of the research being undertaken at the Kyowa laboratories, and it commissioned research to develop the technology needed to enhance octane levels in aviation fuel[30]. Butanol was identified as the starting material to be used for this purpose. The Godo Shusei Company, based primarity on Hokkaido, was selected as a location for the production of butanol. The ABE fermentation technology developed by the Hokkaido Institute of Industrial Technology was transferred to Kyowa Chemical Research Laboratory and, in conjunction with Godo Shusei, an ABE fermentation plant was built at their factory site in Asahikawa in 1938. This plant used both potatoes and corn as the major fermentation substrate.

In 1939 another ABE fermentation plant, associated with these companies, was established in Hachinoe in the Aomori Prefecture located on the northern tip of Honshu. This plant had originally been set up as an alcohol production facility, as part of a government initiative to foster business activities in this poor rural area. The Hachinoe plant consisted of 12 × 90 m^3 fermenters, and was converted to produce acetone and butanol in 1939. This plant also used potatoes and sweet potatoes as the fermentation substrate and produced around 1,500 tons of solvents per year. The former Tohoku alcohol factory was taken over by the main share holders, Godo Shusei, Godo Kyoei, Kyowa Kogyo and Dainihon Shurui. In 1939 Godo Shusei opened a branch in Tokyo, and set up a sales company that secured a contract to provide the Japanese Army with butanol and acetone.

There are also reports of another industrial ABE plant being built around this time situated in the vicinity of Nagoya. This plant was constructed some time after 1939, under the direction of the Kyowa company, and originally consisted 10 × 90 m^3 fermenters. A further 10 fermenters were added later and the solvent recovered by continuous distillation equipment with a capacity of 450 m^3 per day. This could have been the plant located at Ogaki, that was reported to be responsible for around 8% of Japan's total production of butanol later in the war. It appears that there was a second butanol plant located at Yokkaichi near Nagoya and operated by the Dainippon Shurui Company (previously the Ajinomoto Company), that began production in 1942. This is probably the plant that supplied butanol to the Japanese Navy's Second Fuel Depot in Yokkaichi. Takara Shuzo, the third alcohol production company in the Kyowa consortium, was based in Kyoto and might have been associated with the initial ABE plant located near Osaka. Japan also built an ABE fermentation plant at Yanji in Manchuria, close to the border with Korea, which mainly used corn as the substrate.

9.7 Japan's acquisition of US aviation fuel technology

Long before the Japanese attack on Pearl Harbour in December 1941, Japan had begun making preparations for war with the US[14]. A decade before, the Japanese started gathering strategic information for their government by purchasing technical information from the US, making expert observations of industrial facilities and making detailed enquiries during the course of commercial visits. Prior to the war, these commercial relationships, which involved cartel-like arrangements between Japan and various US companies, attracted little attention. As the prospect of war loomed, Franklin D. Roosevelt's administration began expressing a great deal of concern about the way dictatorial and militaristic countries were utilising US business relationships to build their military and industrial capabilities. However, it wasn't until the middle of October 1942, when the US Senate authorised Senator Kilgore to investigate technological mobilisation, as chair of a special subcommittee of the Military Affairs Committee, that the full extent of the business relationships between Japanese and American companies prior to WWII was exposed. This became known as the Kilgore Committee, and over an extensive series of investigations, hearings and reports, the threat that these international cartels posed was uncovered.

These findings led to the establishment of the Office of War Mobilization in May 1943 to deal with this matter[15].

One of the most startling findings that came to light were the use of business relationships by Japan to further its militaristic policies. From the beginning of the 1930s a constant stream of technical information of considerable military importance had flowed back to Japan as a result of Japanese commercial transactions with US firms. The industries most involved included oil, aircraft, electronics and machine tools. Kilgore pointed out that this occurred at a time when US citizens had been barred from visiting islands mandated to Japan by the League of Nations, to ascertain whether or not they were being fortified. The Japanese utilised cartel-like arrangements, such as the sharing of patent licenses, pooling arrangements and technology exchange, to disseminate this information to companies back in Japan. Likewise, technical information that was purchased and observations and enquiries relating to US technical advances and achievements were also made widely available[15].

Japan desperately required the processes for producing aviation gasoline in order to wage war. This knowledge was acquired to a surprising degree from the Universal Oil Products Company in the US. This company had a large research centre at Riverside, Illinois, which employed some 600 technical specialists who were experts in oil refining. The company was actively involved in commercial applications that served a worldwide clientele, and had developed over 1,100 US patents. Much of this information was made available through the normal scientific processes of dissemination. In 1928, financial interests in Japan established the Japan Gasoline Company as the mechanism to acquire Universal technology for the Dobbs process for petroleum cracking utilising heat and pressure. Later that year Japan Gasoline signed a contract with Universal which gave access to the technology, along with any future advances, for which it paid $1,000,000. The technology proved to be unprofitable, and by 1938 only one plant had been built in Japan. However, Japan Gasoline maintained a strong interest in Universal because of their activities in developing new processes for the production of iso-octane. The new developments for the manufacture of aviation gasoline involved two processes, one for the production of the gasoline base stock, and the other for the production of iso-octane that could then be blended with TEL. In 1937 the Japanese made it clearly known to Universal that they were keen to acquire both the base stock production process and the iso-octane process, including the information for the manufacture of the catalysts that were required. At the time most major oil companies were engaged in research in this area. However, catalytic cracking, the alkylation process and the methodology for the production of the catalysts, were all new technologies that were not yet in general usage in 1938. Universal was also made fully aware that the Japanese wished to acquire these processes for the manufacture of aviation fuel for their naval and army air forces. They also knew that all of the refineries under discussion were essentially under government control, and were aware that Japan was also actively working to develop their own polymerisation processes[15].

Understandably, the Japanese government was very interested in the deal with Universal, and the possibility that Japan Gasoline might acquire these processes was of considerable interest to other Japanese companies and organisations. The Japanese considered Universal to be the best oil research institute in the world. The head of the Japanese Fuel Bureau visited the Universal's Riverside Laboratory and pressure was put on Japan Gasoline to close the pending deal. He also made it known that he was willing to help Japan Gasoline become the vehicle to disseminate Universal's technology within Japan. Universal officials were well aware they could be working at cross purposes with the US national interest, and were in a position where the US Government could express strong dissatisfaction with Universal for making these important technologies available to Japan. Negotiations were begun in 1937, and two draft agreements between Japan Gasoline and Universal were in place by August 1938. The first agreement enabled Japan Gasoline to obtain the processes for iso-octane. The second option permitted the Japanese to acquire the rights to all of Universal's processes in the entire petroleum field, including isomerisation, dehydrogenation, cyclisation and alkylation. This agreement included all processes that might have been developed up until the end of 1946. Formal agreements were drawn up in October 1938 that gave Japan Gasoline patent rights as well as access to all of Universal's unpatented technical knowledge and experience. The first agreement was executed following the payment of

$600,000 to Universal. The second agreement was not signed because the Japanese wanted it extended to 1953. On the basis of the first agreement, Universal began to design the iso-octane units to be used by Mitsubishi Oil Company and Nippon Oil, which had agreements with Japan Gasoline. Not only were these designs delivered to Japan in 1939 but Universal engineers were dispatched to help build the plants. Prior to the signing of the agreements, Universal informed the War and State Departments of the pending transactions but indicated that the Japanese could probably secure this knowledge through technical publications, patents and inspection of the plants of competitors outside the US. The War Department expressed no objection. In July 1939 the Japan Gasoline Company exercised option two of its agreement, with a payment of $100,000. This agreement allowed the company's technical staff to visit Universal's laboratories and other research facilities to obtain all the necessary technical know-how. Two months after the option was activated, a group of 35 Japanese technical representatives arrived to spend four months in Universal's facilities. These included petroleum, mining and chemical representatives, technical staff from the Japanese Army and Navy and other Japanese scientists with no contractual relationships with Japan Gasoline. This arrangement essentially functioned as a Japanese government monopoly, providing anyone in Japan access to this information. The technical staff of Universal cooperated fully on providing information for the production of aviation fuel. They conducted lectures and workshops on a variety of subjects, and even reviewed the notes of the Japanese to make sure they had understood the material. Universal also carried out an extensive programme of research in catalytic cracking and hydro-transforming in order to arrive at a type of process which would best meet the Japanese requirements. They also provided the Japanese with research data and test results from pilot plants, along with comprehensive information on catalysts. They then set about designing the plants for the Japanese. This work continued until restrictions were imposed by the US government. Even then, Universal were confident that due to the thoroughness of the technology-transfer activities, the Japanese would be capable of proceeding on their own. Universal considered that the information they had given to the Japanese was more complete than any information on these processes they had supplied to anybody else[15].

In response to the continued Japanese aggression in the Far East, the US government began to apply economic sanctions against Japan. Towards the end of 1938 a 'moral embargo' was placed on the supply of aircraft, engines, parts, armaments, bombs and torpedoes. The embargo did not carry legal restrictions. By December 1939 the restrictions had been expanded to include molybdenum, aluminium and information on the production of high quality aviation gasoline. As a consequence, Universal were obliged to terminate their technical assistance and informed Japan Gasoline that the 1938 agreements would have to be suspended. In June 1940 Japan Gasoline filed suit against Universal, seeking reinstatement of the contracts or payment of $100,000,000. The case was postponed due to the advent of the war. This setback did not deter the Japanese, who, through appropriate internal agreements, were able to proceed with the planning and construction of five catalytic cracking plants, with a total daily capacity of 15,000 barrels. Each plant also included units to manufacture iso-octane for blending as well as five independent iso-octane units with a 650 barrel daily capacity[15].

At the time that Universal was making this information available to the Japanese, its American partners were having difficulty securing the same kind of information. Shell, which owned 50% of Universal's voting stock and was a licensee for all Universal patents through to 1947, complained that information provided to them in 1938 concerning cracking and catalytic reforming was incomplete. Shell requested that Universal divulge the full details of their research relating to the design and operation of these processes. In December 1941 Universal responded that Shell was not entitled to this information. Likewise, the Standard of Indiana that held over $300,000 of Universal stock, and was also a paid-up licensee through to December 1947, was denied access to reports and other information that had been widely distributed to the Japanese. The Universal board of directors supported this position, and ruled that under these agreements the US firms were obliged to purchase their catalyst, while the Japanese firms were not. The Japanese, however, did buy a year's supply of the catalyst from Universal and thus received the rights to the catalytic technology. Later, fearing that this kind of information would find its way into the hands of the Kilgore Committee, J. G. Alther, vice president and general manager of Universal, begged Justice Department officials not to release this information[15].

The Mitsubishi Oil Company was another Japanese organisation that developed important strategic relationships with US companies. This company was formed in 1930 as a partnership between the Mitsubishi Trading Company and the Associated Oil Company in the US and subsequently became known as the Tide Water Associated Company. The two partners planned a joint venture in occupied China, and William F. Humphrey, Tide Water's president, was handsomely remunerated by Mitsubishi. Humphrey was a staunch supporter and advocate for Japanese business interests and he lobbied extensively on their behalf in Washington. In 1940 Mitsubishi requested that Humphrey assist in ensuring that US petroleum supplies to Japan were maintained. After Roosevelt's proclamation forbidding the export of plans, designs and information that could be used in the production of high octane aviation gasoline, Tide Water was placed in the position of having to deny Mitsubishi's request for reference fuel, which was to be utilised for the measurement of octane ratings. Several months after the introduction of the moral embargo, the Cooper Oil Company supplied Mitsubishi with technical information concerning the production of neo-hexane and other alkylates, which were being developed as substitutes for iso-octane. Cooper also supplied Mitsubishi with confidential information on the 92 octane gasoline process from Maritime Oil Company. Cooper officials wrote that these reports were being provided in complete confidence, and should not be disclosed, as it would place the company in a very difficult position if this information ever leaked out. Even after a full embargo, forbidding the sending abroad of information regarding high-octane gasoline and other items, was in place, Orchard Liske, a consulting petroleum technologist, provided Mitsubishi with strategic information of military value regarding the production of toluene for use in explosives. This information covered the pyrolytic treatment of petroleum to yield toluene that would allow Japanese engineers to replicate the process. He informed Mitsubishi that this information did not appear to be specified in law – so long as actual refinery and process designs were not provided. Mitsubishi also used its extensive business contacts in the US to pass strategic military information back to Japan. This included detailed information on the production and supply of 100-octane aviation gasoline to US naval air stations and army airfields, as well as shipments of petroleum and fuel oil going to Pearl Harbour. Later investigations by the Justice Department revealed that Mitsubishi, which had been the largest Japanese company operating in the US, had functioned as an effective spying agency for the Japanese government. It is not surprising that the company destroyed all its records in December 1941. In another case, the Ocon Petroleum Processes Corporation entered into discussions with the Okura Company of Japan regarding access to their catalytic naphtha reforming process. As the moral embargo had been proclaimed three months earlier, the company was afraid to demonstrate its process in the US but arranged a demonstration of this process at their plant in Mexico[15].

The Justice Department's Board of Economic Warfare cooperated fully with the Kilgore committee, and the possibility that information gleaned through Justice Department investigations might get to Kilgore struck fear in the hearts of business executives who did not want their questionable activities exposed. Kilgore was forced to conclude that numerous American companies had been willing to provide strategic information of military value to Japanese business – for profit and to the detriment of the US – and he wondered why, with the kind of pressure brought to bear in Washington by companies like Tide Water, the US was surprised by the attack on Pearl Harbour[15].

9.8 Developments leading to use of biobutanol for iso-octane production

In 1937 the Japanese government passed the Alcohol Monopoly Law which empowered it to take appropriate action to ensure the maintenance of a stable supply of liquid fuels. The government selected the Kyowa Chemical Research Laboratory in Tokyo to undertake research to develop technology required to enhance octane levels in aviation fuel[30]. The government contract was a turning point for the newly formed research facility. At this time Benzaburo Kato was the director, and under his leadership the laboratory focused on developing the chemical technology for the synthesis of iso-octane by alkylation and polymerisation[30]. Independently, around 1942, the Japanese army initiated research at the Army Fuel Research Centre in Tokyo on processes to produce iso-octane.

The US moral embargo implemented in 1939 was largely responsible for preventing Japan from gaining full access to American 'know-how' relating to the production of iso-octane and catalytic cracking. Despite this barrier, Japan did develop a number of processes for the production of iso-octane for aviation fuel. However, the production of iso-octane remained low due to an inadequate supply of C4 hydrocarbons and a limited refining capacity. Post-war assessment indicated that annual peak production only amounted to around 80,000 barrels. This was mostly used for experimental purposes, and very little appears to have been available to upgrade aviation fuel for combat aircraft until very late in the war. A number of processes for the production of iso-octane were developed. In one process iso-octane was synthesised from acetylene that was produced from calcium carbide. A number of companies based in Manchuria were involved in the establishment of a chemical process using coal and lime to produce acetylene. Neo-hexane was also investigated as an enhancer of octane levels in aviation fuels, as acetylene could also be used in a synthetic chemical reaction to produce acetone that could then be converted to neo-hexane. A drawback to these processes was that they were all very energy intensive[28].

The production of iso-octane was also achieved by the polymerisation of n-butanol fermentation using Japanese acidic clay as a catalyst. Butanol was converted to butene, which was dimerised to form octanol, and then hydrogenated to iso-octane. A process to synthesise iso-octane produced from n-butanol and butane obtained from casing-head gas also appears to have been developed. Research on producing butene from the lower hydrocarbons, methane, ethane, propane, ethylene and propylene was a further aspect that was investigated. Butanol produced by fermentation was identified as the primary feedstock to be used for the production of iso-octane. The industrial ABE fermentation also produced either acetone or isopropanol that could be used as raw material for the production of neo-hexane. The need to produce as much butanol as possible led to the expansion of the industrial ABE fermentation process.

An obvious choice for the expansion of the ABE fermentation industry was on Japan's colony of Taiwan.

9.9 Reasons for the establishment of a major butanol fermentation industry in Taiwan

The defeat of the Manchu Empire by the Japanese in 1895 in the First Sino–Japanese War led to Taiwan being ceded to Japan in perpetuity as part of the post-war settlement package. It was administered as a Japanese colony from 1895 to 1945. Taiwan produced rice, sugar and a range of other agricultural products, and had significant coal deposits along with some oil and natural gas deposits. However, although Taiwan had been part of China since the late 1600s there had been very little development. The Japanese immediately set about demonstrating what could be achieved with their first major colonial holding. The Japanese rulers were instrumental in bringing about the rapid modernisation and industrialisation of the island. Japanese became the official language and Shinto became the official religion. The public school system was revised, and Japanese Imperial education was implemented. By 1938 there were in excess of 300,000 Japanese settlers in Taiwan, and much of the island had been transformed to look like Japan, with modern cities, housing, schools, hospitals, industries and farms. During the 50 years of occupation the Japanese built up a tightly integrated industrial base. To assist the development of industry an excellent communications network was built, consisting of a double tracked rail system (Fig. 9.1), along with concrete roads and bridges. A number of large, mostly artificial, ports were built in the north, east and south. Energy was supplied by hydroelectric power and coal, and hydroelectric power was used to refine aluminium. There were eight small oil fields on Taiwan, most associated with natural gas deposits. By the time war broke out there were three oil refineries and eight casing-head gasoline plants in operation, along with three ocean handling terminals. The size of the petroleum industry on Taiwan was small, with an annual output amounting to just 110,000 barrels in 1938 –equivalent to less than one day's production by the US industry[27].

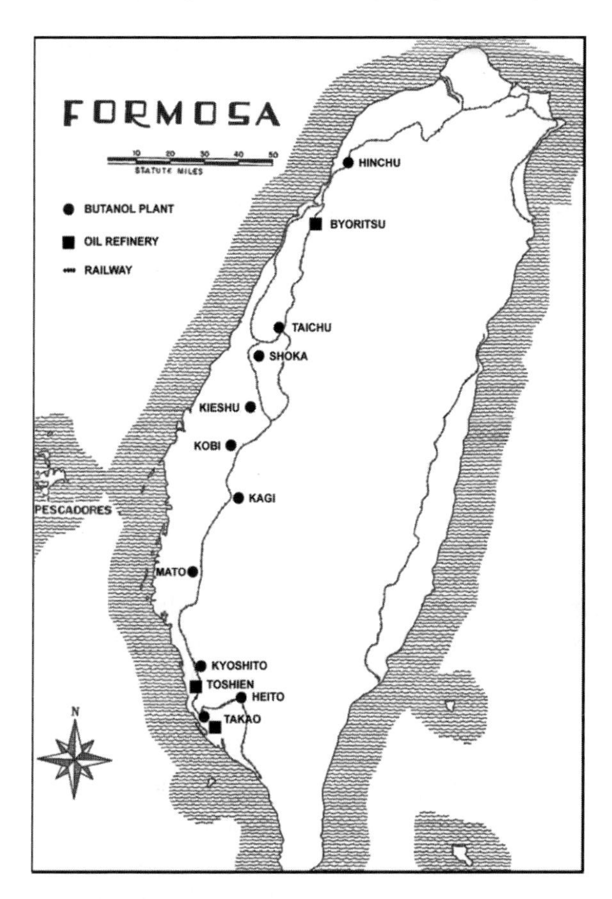

Fig. 9.1: USAAF (US Army Air Force) map of Taiwan modified to show the location of butanol plants and oil refineries.

During the period of Japan's occupation both rice and sugar cane production was greatly increased, and Taiwan became a major exporter. An extensive sugar industry was developed, which produced around 17,000,000 tons of sugar annually. By 1939 Taiwan had become the seventh largest sugar producer in the world. The industry was under private, mostly Japanese, ownership, but the Japanese authorities exerted a significant degree of control over the growing of sugar cane and rice. In 1927 there were 11 sugar companies operating in Taiwan. By the early 1940s mergers had reduced this number to four main companies, with a total of around 30 large sugar mills/refineries[31]. From 1927 through to the early 1940s sugar production was tripled. Alcohol plants were located adjacent to most sugar factories and refineries, and both sugar- and starch-based agricultural crops were used for ethanol production. By 1939 ethanol had become a well established export industry, and the ethanol fermentation process was considered to be a mature technology.

In addition to its natural resources and well developed infrastructure, there were a number of other factors that made Taiwan an obvious choice for the establishment of the butanol fermentation process. A central Research Institute had been established in 1921 that included the Southern Resources Institute in Sciences. Taipei Imperial University had extensive well established linkages with Japanese universities. Petroleum research was initiated in 1936, along with research into ethanol fermentation[32].

Taiwan had also been developed as a major Japanese military base to act as a springboard for the Japanese colonial expansion into Southeast Asia and the South Pacific. As part of war planning an elaborate air defence network that included around 50 airfields was constructed. These facilities were used mainly for pilot training, but combat missions were also flown from the island.

The Imperial Japanese Navy also operated extensively out of the ports around the island. The Army's 'South Strike Group' was based in Taiwan, and many military instillations and training bases were built. Tens of thousands of Taiwanese also served in the Japanese military. The Sixth Navy Fuel Depot/Factory became the official Japanese Navy name for the group of refining facilities and factories established in Taiwan to produce aviation fuel.

9.10 The main butanol plant at Kagi in Taiwan

The city of Kagi (Chiayi) in southern Taiwan was selected as the site for the initial butanol plant because it was a regional centre for the production of sweet potatoes (Figs. 9.1, 9.2, 9.3). The plant was constructed in 1939 specifically to manufacture butanol, acetone and ethanol by fermentation, and the first test runs were begun in June 1939[33,34]. Initially the factory was owned and operated by the Taiwan Development Corporation/Company that was under the control of the

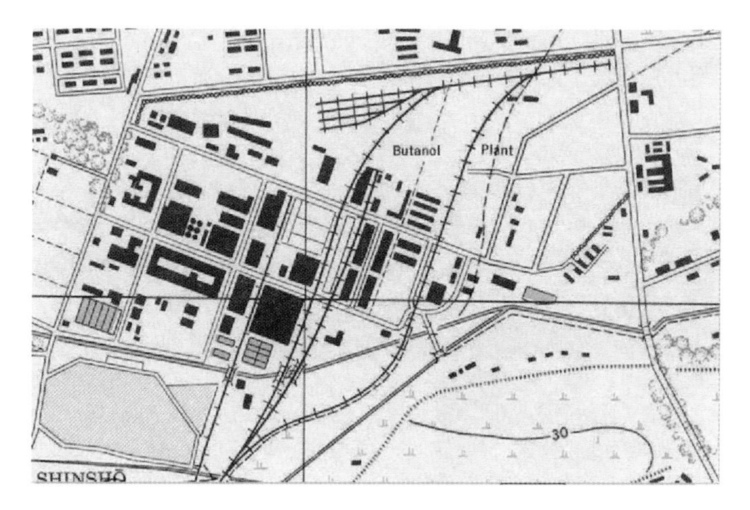

Fig. 9.2: Site map of the butanol plant at Kagi from1944.

Government National Resources Bureau. The Corporation was a semi-public entity established in the 1930s by Japan, and sponsored by the Taiwan Government General to support its strategic aims of economic expansion into Southeast Asia and the Pacific islands. In 1942 a joint arrangement was entered into with Japanese Navy, and the name of the plant was changed to the Taiwan Development Chemical Industry Company.

A USAAF intelligence report based on aerial reconnaissance photographs and maps dated April 1944 gave an overview of the Kagi butanol factory at this time (Fig. 9.2). The butanol plant was located at the southern edge of the city, and occupied an area of around 46 hectares. The factory was situated adjacent to a canal that drew water from a dam located on a major river nearby. The plant was served by a railway spur that branched off the main line and entered the factory from the south west. The plant was also serviced by a narrow-gauge railway used to transport raw materials from the surrounding areas. A large coal storage receiving area was located adjacent to the rail tracks. Next to this was a large heat unit with a smokestack. A steam unit consisting of two buildings with a smokestack between them was also located just south of the main plant (Fig. 9.3). A transformer was located to the south east of the main plant, connected by a subsidiary power line to the power grid

Fig. 9.3: The main butanol plant at Kagi. A. The newly complete plant in June 1939. B. View of the plant some time after 1939. C. Aerial view of the plant showing the main fermenter building taken during an air raid during 1945. D. The plant after it was 90% destroyed at the end of WWII.

running from the north to the south east of Kagi. There were four large receiving warehouses and six smaller warehouses located at the centre section of the plant adjacent to the rail tracks. The initial stage of the of the process was housed in a very large building which was thought to contain the mills used for processing the raw materials and the first part of the sterile mash preparation (Fig. 9.4). A second large building was connected to this building, and was thought to house the separators, mash cooker-coolers and the final sterile mash holding vessels. A long rectangular-shaped building enclosing an open rectangular yard contained the fermentation tanks, which were approximately 5 m in diameter. Several miscellaneous buildings and two large storage tanks that might have contained either by-products or raw materials, such as molasses, were also located in this area. Another large composite

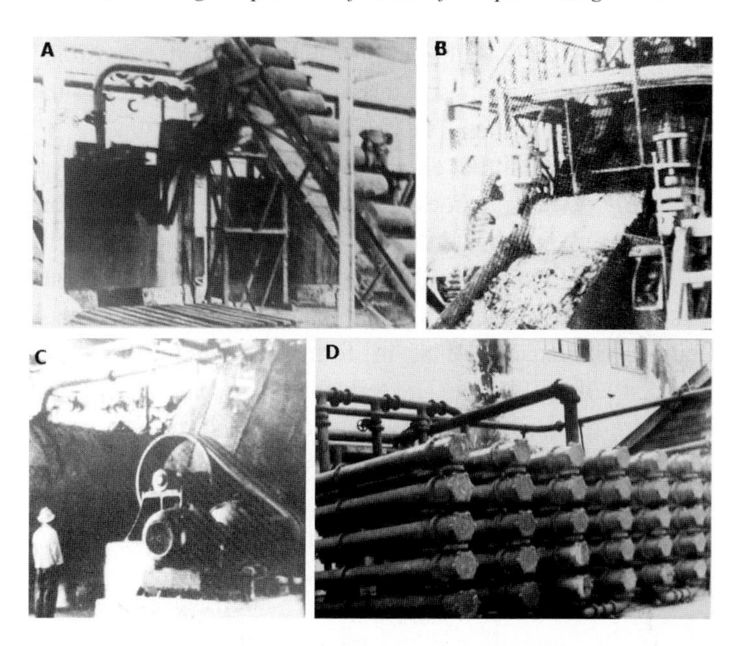

Fig. 9.4: The butanol plant at Kagi showing the substrate preparation equipment. A. Lifting gear. B. Roller mill. C. Mixing tanks. D. Mash preparation.

building contained the distillation fractionating columns, and the steam was supplied by the large heat unit located close by. A smaller building with six tanks of around 10 m in diameter, located nearby, was thought to be used for storage of the finished products. An earth scar, indicating the probable presence of a pipeline, extended from a small building located in the centre of the plant to a point 1,500 m to the east, ending in a pond plus three small buildings that probably handled the effluent from the factory. Approximately 17 office buildings, laboratories and other miscellaneous buildings were located in the western section of the factory. The report also states that a second production unit consisting of three new buildings, similar in size, shape and arrangement to the three main production plant buildings, were under construction on the northern side of the existing plant. It was concluded that when this new production unit was completed it would double the production capacity of the plant. The area east of the existing plant also showed evidence of extensive building activity that included extensions to the rail yard, additional warehouses and several new structures of unknown function that were in various stages of construction (Fig. 9.2).

The plant employed the classic ABE batch fermentation process, utilising mainly sweet potatoes, along with some corn, sorghum and other starch-based raw materials, using a derivation of the original Weizmann strain and other similar Japanese isolates. At full capacity the plant consisted of 92 fermenters, each with a capacity of 170 m^3 plus six fermenters with a capacity of 85 m^3 fermenters giving the plant a total reactor capacity of 16,150 m^3. The fermenters were inoculated from eight mid-sized pre-fermenters, using standard laboratory seed stage propagation (Fig. 9.5). The raw material was brought in from the surrounding countryside by way of a narrow-gauge rail network. The substrate was them milled and prepared in four mixing tanks with a

Fig. 9.5: The butanol plant at Kagi showing the fermentation equipment. A. Pre-fermenters. B. View of fermenters from upper level. C. Side view of the main fermenters. D. Distillation columns.

volume of 17 m^3 and then processed by way of 22 mash cookers with a capacity of 34 m^3 and another two mash cookers with a capacity of 13 m^3 (Fig. 9.5). The plant also had one pressure filtration unit of 900 m^3. The solvent was recovered and separated by distillation. In full production the factory was reported to have employed 3,200 people, and was the biggest of its type in the Far East. One-fifth of the population of Kagi was said to have depended on it for employment, and much of the country was devoted to growing sweet potatoes for the factory.

When the plant began operation in 1939 the initial production was less than 200 tons of butanol. This increased to almost 1,500 tons after merger with the Navy. During 1943 and 1944, production was increased to around 5,500 tons a year until 1945. Although the main fermentation continued to utilise sweet potatoes, in 1943 the plant was also adapted to using sugar, and in 1944 switched to using cane juice and syrup. The plant did not have a sugar milling facility, so the cane-base substrates were brought in by rail. The first USAAF air raids commenced early in 1945. During the last months of the war the plant was converted to producing ethanol for use as a motor fuel.

9.11 Japan's fuel situation at the outbreak of WWII

The Japanese invasion of China in June 1938 placed the imports of oil and aviation fuel from the US under threat. Policy makers in the US balked at continuing to sell fuel to the Japanese to enable the Imperial Army to run roughshod over the Asian mainland. However, the US remained cautious, as pre-war US Navy analysis had concluded that an oil embargo was likely to result in an immediate attack by Japan on Malaya and the Netherlands East Indies (NEI). This would inevitably involve the US in an early war in the Pacific. Meanwhile, the Japanese continued to stockpile as much Californian and Mexican crude oil as possible and even attempted to buy outright one potentially oil-rich area of Mexico. In 1940 Japan sent a mission to Java seeking to buy 3,500,000 of tons crude oil as well as aviation fuel. The conference with the Dutch government went badly and the oil companies were also uncooperative. The final sale that was made consisted of 1,000,000 tons of refined oil and 760,000 tons of crude oil. No aviation fuel was supplied, as this had already been contracted to the allies[35]. By the end of 1941 Japan had a stockpile of around 60,000,000 barrels of crude oil[19]. Japan's own annual production amounted to 2,500,000 barrels, while its annual requirement amounted to around 30,000,000 barrels. In 1941 Japan's refining capacity was estimated to be around 35,000,000 barrels. Japan's fuel stocks of petroleum and aviation fuel were being rapidly drained by its war with China. It was estimated that if war were to break out in the Pacific, existing stocks would only last for two years. The Army argued that the only option was to secure the oil from the NEI oil complex in Borneo, Java and Sumatra, and this meant it would be necessary to go to war to achieve this. They also advocated that, since war was inevitable, there should be no further delay. They proposed that an attack should be launched immediately, while abundant fuel stocks were still available. Both the Japanese government and the Navy disagreed. The former were of the view that they could not win a war against the US, while the latter believed this could be achieved, provided that the war did not exceed two years in duration.

The drive for oil led Japan into the first oil paradox of the Pacific War. The Japanese, fearful of a total US oil embargo, sought to gain alternative sources by gaining control of oil-producing territories in the East Indies. However, it was precisely this expansionist policy which precipitated the embargo. A Japanese initiative that forced the closure of the Yunnan Railway in July 1940, led to a US trade embargo on the sale of aviation fuel to Japan. When a total trade embargo against Japan was finally put in place in October 1941, US military planners accepted that war in the Pacific had become inevitable. Intercepted and deciphered radio traffic led the US to believe the Japanese would head straight for the sources of oil in Indonesia and Malaya. The fields in the NEI yielded 170,000 barrels of crude a day, and were only lightly defended. The possibility that the Japanese might first make a major effort to put the US Navy out of action did not enter most analysts' minds. The attack on Pearl Harbour was planned as a surprise military action to be conducted by the Imperial Japanese Navy against the US naval base in Hawaii. The attack was intended as a preventive strike to keep the US Pacific Fleet from influencing the war that the Japanese Empire had planned in Southeast Asia, against Britain and the Netherlands, as well as against the US in the Philippines. The attack that was launched against the naval base on the morning of December 7, 1941 was carried out by a total of 353 Japanese aircraft in two waves launched from six aircraft carriers. The surprise attack caused over 3,000 casualties and sank or damaged many ships including all eight battleships anchored in the harbour. Though the attack severely incapacitated the US Pacific fleet, it united a previously divided US and committed the nation to war. The following day the US declared war on Japan.

In December 1941 Japanese forces invaded Borneo, followed by Sumatra and Timor in February 1942 and Java in March 1942. The Dutch, British, Australian and American forces surrendered in March 1942. This gave the Japanese almost total control over the Pacific theatre for a short period, and they were able to secure most of the oil in the NEI, despite allied attempts to destroy those facilities. Technicians accompanied invading troops to repair damage intentionally inflicted by the previous operators. These preparations paid off, and by the end of 1943 they were able to restore crude oil production to 76% of its 1940 level. Even more importantly, they were able to restore refined products to 45% of 1940 levels, reducing the load on tankers and home island refining facilities. For a time Japan was able to do what Germany was never able to do — achieve independence with respect to oil. However, within a short time the Japanese began to experience major difficulties that rendered their possession of the NEI oil fields of limited value. This second oil paradox of the war in the Pacific occurred because the Japanese did not have a sufficient number of tankers to ship the desperately needed oil and gasoline to the home islands, or to service their many outposts spread across the vast new Japanese Empire. The Navy experienced increasing difficulty in protecting the tankers they did have from the attacks by submarines, aircraft and surface ships. The oil had to travel thousands of miles to get from Balikpapan in Borneo to home ports in Japan, and US submarines were lying in wait all along the route. One of the most important ship sinkings of the war occurred when a US submarine sank the Taiyo Maru in the summer of 1942. The ship had over 1,000 Japanese petroleum experts and technicians on board, and was headed for the NEI to spur on oil production. A total of 780 personnel perished in the attack. By the end of the war, 110 Japanese tankers had fallen victims to allied submarines. Protective measures by the Japanese proved to be ineffectual, as US cryptanalysts had broken the Japanese naval code, and were fully informed of tanker schedules and cargoes. The Japanese Navy was also slow to respond, and did not establish convoys for the vulnerable tankers until 1943.

9.12 The development of a sugar-based butanol fermentation process

With the advent of the war in the Pacific, increased production of butanol was given a high priority. Although large volumes of sugar cane-based substrates were available in Taiwan, molasses and cane juice could not be used due to the lack of suitable saccharolytic strains. The rapid advances made by the Japanese into the Philippines and East Indies meant that additional sugar-based material became even more widely available. As a result, a process to enable the utilisation of sugar cane-based substrates for the production of butanol was considered to be of major importance.

In December 1941, on the eve of the Pacific war, a wide ranging planning forum was held in Taiwan to review research undertaken during the 1930s. The main purpose of the forum was to prioritise research themes to be developed during the 1940s[32]. This policy meeting was attended by high ranking officials from the Japanese administration and ministries, research institutes and the Taipei Imperial University. As a result, a number of new programmes in agriculture and industry were identified. A programme to develop a butanol fermentation process using sugar cane as the substrate was amongst the most significant and was given top priority. The Southern Research Institute coordinated the programme involving various institutions in Taiwan, such as the Industrial Research Institute, the Sugar Research Institute, the Sugar Research Society, the Taipei Imperial University and the Taiwan Monopoly Board. A number of universities in Japan also embarked on an intensive programme to isolate and develop new strains. Success was finally achieved when the Southern Research Institute isolated a strain that proved suitable for the purposes[32]. This strain, initially referred to as the Baba strain, was later described in the published literature as

Clostridium toanum[36]. Another strain isolated in Japan around this time was also described in the Japanese scientific literature in 1943 as *Clostridium butanologeum*[37].

A description of *C. toanum,* based on this original description[36] was included in the seventh edition of Bergey's Manual of Determinative Bacteriology[38]. The cell morphology was described as large rods (1.04–5.47 μm) occurring singly or in short chains, becoming spindle shaped (1.9–7.35 μm) after 48 hours. The cells were actively motile in young cultures and were granulose-positive and Gram-positive, becoming Gram-negative in old cultures. The cells were encapsulated and the spores were ovoid, located sub terminally or centrally. The strain was an obligate anaerobe with an optimum temperature of 33°C. Growth occurred between pH 6.0 and 7.0, with optimum growth at around pH 6.2. Optimum conditions for fermentation were between 35° and 37°C, with a pH between 5.8 and 6.5. Solvents were produced in the ratio of around 53:40:5:2 butanol : isopropanol : acetone : ethanol, with some butyric acid also being produced. A sugar concentration of 70 g/l produced around 21 g/l solvents with 11–12 g/l butanol at a yield of around 30%. The new strain did not liquefy gelatine and did not produce endol or hydrogen sulphide, but nitrites were produced from nitrates. Milk was coagulated but the organism grew poorly on tryptone and meat infusion media. Glucose, arabinose, fructose mannose, galactose, sucrose, maltose, trehalose, alpha-methyl-glucoside, dextrin, soluble starch and glycogen were all vigorously fermented. Raffinose, salicin, mannitol and beta-cellobiose were fermented less vigorously and xylose, lactose, pectin, calcium lactate and amygdalin were only weakly fermented. This strain did not ferment glycerol, adonitol, dulcitol, sorbitol inositol, quercitol, rhamnose or inulin[38]. From the description provided it seems likely that this strain belonged to the saccharolytic, solvent-producing clostridia related to the *Clostridium beijerinckii* cluster. None of the major international culture collections list this strain amongst their holdings[39].

The new process using cane juice, cane syrup and molasses employed steam sterilisation, and the Houjiaruma fermentation process, with solvent recovery by distillation. The optimum sugar concentration used for the process was 7–7.3% and the substrate composition used for the fermentation was mixed in the ratio of 100 tons sugar, 7 tons rice bran, 5 tons calcium carbonate, with a set sugar concentration of around 7%. The steam needed to process 100 tons of sugar required between 100 and 200 tons of coal. One hundred tons of sugar substrate produced around 16 tons butanol (53.3%), 12 tons isopropanol (40%), 1.5 tons acetone (5%) and 0.5 tons ethanol (1.7%). The total solvents amounted to around 30 tons, representing a yield of around 30%. Fermentation times for the new sugar-based process were reduced to almost half, and the batch fermentations were completed in 30–40 hours[32].

Under the direction of the Taiwan Monopoly Bureau, a programme for the mass production of butanol from sugar cane was implemented. A number of new butanol plants were constructed during the years 1942–1944 (Fig. 9.1). The four main sugar companies — Taiwan Sugar, Meiji Sugar, Nitto Development and Ensuiko — established a cooperative unit called Taiwan Synthetic Fuels to manufacture iso-octane from butanol[26]. Various wartime US intelligence reports identify the existence of at least ten butanol plants situated at Heito, Hinchu, Kagi, Kieshu, Kobi, Kyoshito, Mato, Shoka, Takoa and Taichu. Most of these were associated with sugar mills, sugar refineries, ethanol plants or oil refineries that were in operation at this time (Figs. 9.1, 9.6). Another report states that the government responded by converting more than half of the country's 30 giant alcohol production plants to produce butanol for use as aircraft fuel, solvents and lubricants. This suggests that there could have been more than 15 butanol plants in operation at this time. The plant at Kagi was reported to have the largest capacity and another plant, also operated by the Taiwan Development Company, located close to the oil refinery and deep-water port in Takao, was reported to have a capacity of about two-thirds that of the Kagi plant[26].

A new ABE plant was also built on Hokkaido at Shimizu-cho in the Tokachi district. This plant was established in 1943 at the Meiji sugar company, to utilise sugar beet molasses as the substrate, and became operational in 1944. Another reference indicates that seaweed might also have been used as a substrate on Hokkaido[40]. Other reports mention that at a later stage of the war the Japanese also built an ABE plant in Java[41] and another in the Philippines[42]. The Godo Shusei company was apparently involved with these developments. From the sketchy

Fig. 9.6: Aerial views of some of the sugar factories with associated butanol plants taken during USAAF bombing raids in 1945. A. The plant at Heito. B. The plant at Kyoshito. C. The plant at Mato/Ensui. D. The plant at Shoka.

information available it seemed that a number of difficulties and teething problems were encountered, and operations at both plants, intended to utilise sugar cane, were short lived, so little or no butanol was produced.

9.13 The role of the Japanese Navy fuel depots in butanol and iso-octane production

The Japanese Navy did not rely on industry for petroleum research and refining. Instead it established its own in-house facilities, which were designated as naval fuel depots/factories. These comprised six depots associated with the various designated naval districts. The first depot consisted of the Ofuna fuel and lubricant research centre that was built near Tokyo just before the war. This was one of the world largest fuel and research institutions, consisting of 70 modern buildings staffed by about 3,200 personnel. The second depot consisted of the Yokkaichi petroleum refining facility, built in 1940, with a capacity of 17,000 barrels per day and the third depot consisted of the Tokuyama petroleum refining facility, built in 1920, with a capacity of 9,500 barrels per day. These two facilities were two of the largest refineries in Japan, and used technology that was largely of US origin, with some Japanese developments. Together they made up 25% of Japan's refining capacity. The fourth and fifth depots consisted of the Shinbara mining and carbonisation facility on Kyushu and the Heijo Coal mining and bunkering facility in Korea. The Navy also operated other fuel facilities at Samarinda and Balikpan in Borneo. The sixth naval depot/factory was located on Taiwan and consisted of three separate facilities.

The development of the butanol and iso-octane production process led to the building of a commercial-scale plant, erected for the Navy, at the second naval fuel depot at Yokkaichi. This industrial facility, named the Japan Butanol Company, was apparently operated by Kyowa Sangyo Company and began producing butanol in 1942. This ABE fermentation plant appears to have been associated with a facility for the production of iso-octane manufactured by a catalytic fixed bed process utilising Japanese acid clay catalyst.

At the beginning of 1942 the Navy entered into a joint arrangement with the ABE factory in Kagi and this facility became known as the Taiwan Development Chemical Industry Company. After the war against Japan came to an end the US Navy sent a technical mission to Japan from September 1945 to November 1946. The mission comprised 655 officers and technicians, and approximately 3,500 documents were seized and shipped back to the Washington Document Center and the technical bureaus of the Navy Department. Amongst these were a number of reports on research undertaken in Taiwan by Japanese naval personnel between 1941 and 1944 on the optimisation of the butanol fermentation process[43]. Some of the reports detailed work undertaken by Chem. Eng. Lt. Comdr. T. Umemura, assisted by Chem. Eng. Lieut. M. Takahashi. Other reports covered work undertaken by Mr Y. Takeda and S Himada, Chief Engineer and Engineer from Central Research of the Formosan Government General. These include the assessment of at least

40 new solvent-producing strains isolated from soils, using the Weizmann strain as a control. Research included both the optimisation of the fermentation using sweet potatoes and molasses as substrates, utilising strains previously selected and developed in Japan. Both the 12 KN strain and the Weizmann strain gave similar results on sweet potato. The reports also covered investigations to try to enhance butanol production in 4% and 6% sugar mash substrates, using exposure to X-rays, electric lamps and super-sonics[43].

The sixth naval factory on Taiwan was spread over three different sites (Fig. 9.7). The facility at the port of Tako comprised of a petroleum refinery. In 1942 planning was initiated to build a new butanol fermentation plant associated with a facility for the production of iso-octane at Shinchiku (Hinchu). The planning and building of the butanol plant was completed in 1943 and by 1944 the plant was producing butanol. This plant was said to have a reactor capacity of 10,000 m^3. The second phase of the project to produce iso-octane had still not been completed by the time the plant was destroyed by USAAF bombing in 1945 (Fig. 9.7). The iso-butanol facility was reported to have included an electro-arc unit and a gas washing plant, presumably to be used for the processing of C4 containing casing-head gas from the oilfields located nearby. The third facility of the sixth naval factory was located at Niitaka to develop lubricants from caster oil, rape oil, peanuts and coconuts. There were around 2,500 personnel associated with the sixth naval factory, of these around 1,200 were stationed at Tako, 900 at Shinchiku and 300 at Niitaka.

In 1944 US intelligence estimated that Japan was aiming to produce 75,000 tons of butanol, which would have been sufficient to produce 3,000,000 barrels of high octane aviation fuel. Butanol was used both as source of iso-octane production and was also added directly into aviation gasoline. A 1945 US intelligence report states that Japan had 19 butanol plants of unknown capacity and uncertain location, and it provided estimates of capacity for five plants known to be located in Taiwan as 3,200,000 gallons per annum[25]. Figures included in the report prepared by the US Navy Technical Mission to Japan in 1946 indicated that the production of butanol and acetone more or less doubled every year during the four-year period from 1940 to 1943, but had declined sharply

Fig. 9.7: USAAF bombing targets in 1945. A. A Liberator bomber over the port of Tako — location of the oil refinery and butanol plant of the sixth Japanese depot. B.The bombed ruins of the butanol and iso-octane plant of the Japanese depot in Hinchu. C. A USAAF intelligence photograph identifying the sugar refinery and the ethanol and butanol plants at Kobi.

by the end of 1944. At this time the importation of raw material had largely been halted, and it was surmised that the local production of raw materials had to be diverted to food production in order to prevent starvation. It is interesting to note that serious widespread problems with phage infections were reported to have been encountered around this time. This was referred to as 'sleeping sickness' and it was reported that both the sugar bases and starch-based fermentations were affected at a number of factories[44]. The industrial ABE fermentation process in Japan was also plagued by phage and bacterial infections, and ongoing difficulties were encountered with maintaining sterile conditions and problems with substrate quality.

9.14 The USAAF bombing campaign in Taiwan, 1944–1945

The allied counteroffensive in the Pacific War involved no half-measures. A key aspect of the US Navy's Pacific strategy was an intense campaign against Japanese commercial shipping, with the primary target being to disrupt the importation of oil. Although the blockade, spearheaded by US Navy submarines proved a highly effective means of disrupting Japan's oil supplies, the USAAF leaders opted for the strategic bombing of the Japanese home islands, including fuel facilities, rather than providing support for the naval blockade. To some extent this service parochialism impeded the implementation of a rational overarching US military strategy for the Pacific theatre.

The Japanese were willing to risk everything to defend the Philippines, because the location of these islands made them critical for defending the long shipping lanes running from Borneo and Sumatra to Tokyo. But at Leyte Gulf, with General MacArthur's invasion force still vulnerable to counterattack, the Japanese 2nd Fleet turned back only 40 miles from the beaches as it was considered that the fleet was too short on fuel to risk an attack. Once the Philippines had been retaken the USAAF came within bombing range of Taiwan. At this stage it was known that the Japanese had large-scale facilities for the production of ethanol and butanol, used for the production of aviation fuel, that were based around the widely scattered sugar mills in Taiwan. As a consequence, air attacks against sugar mills and alcohol plants in Taiwan became the most extensive and persistent of any bombing attack on the island with the exception of those targeting the neutralisation of Japanese air power (Figs. 9.3, 9.6, 9.7). Bombing raids were begun on a small scale in January 1945, and reached a peak in May before slackening off in June due to other priorities. Raids ended in July with only a few fighter sorties flown in August. Initially the targets were only lightly defended by anti-aircraft guns, and in the early phase these targets were assigned mainly to medium bombers. The increasing concentration and accuracy of Japanese anti-aircraft defences around these targets led to a shift in tactics, with heavy bombers participating for the first time.

During this period some 30 sugar mills/refineries and alcohol/butanol plants were bombed on a daily basis, except when bad weather prevented flying (Figs. 9.1, 9.3, 9.6, 9.7). These included all of the known plants in Taiwan. After the war the USSBS (US Strategic Bombing Survey) team credited the bombing campaign with the destruction of at least 75% of the plants. The report from the Japanese Governor General Office listed 17 plants as completely destroyed, nine moderately damaged and four slightly damaged. Still further reduction of the enemy's potential fuel supply was attributed to the extensive disruption of rail transportation, including the destruction of rail and road bridges. It was considered that the effects of the intensive bombing resulted in a shift from growing sugar cane to rice, as this was required to keep Taiwan self sufficient in food. Japanese military bases also came in for special attention. As the majority of the strategic targets in Taiwan were situated in the cities and towns, area bombing was frequently employed. The massive destruction resulting from this was aimed at not only destroying facilities and supplies of military importance, along with the destruction of many small industrial units, but would also impose a serious loss of labour, through destruction of housing and municipal services. Out of 11 principal cities and towns the Governor General Office later reported that five had been almost completely destroyed, and four had been 50% destroyed. Hinchu City suffered the greatest bomb damage in Taiwan, based on area. The butanol plant in Hinchu that was part of sixth Navy Fuel Factory, which was believed to have synthesised iso-octane aviation fuel, was completely destroyed (Fig. 9.7). In addition to government buildings, factories and industrial plants, more than 10,000 buildings were destroyed, 16,000 were damaged and around 20,000 were totally or partially burnt. All of the main harbours and most of the smaller harbours were destroyed, blocked or damaged, and all shipping movements were halted. The USAAF Fifth Bomber Command carried the main burden of the attacks, flying 87% of the total sorties, and dropping 98% of the bomb tonnage. Liberator B-24 heavy bombers flew 5,000 sorties (Fig. 9.7), Mitchell B-25 medium bombers flew 1,500 sorties, while Havoc A-20s, Invader A-26s and Dominator B-32s flew just over 200 sorties. A total of 13,804 sorties were flown, during which 15,804 tons of bombs and 62,454 gallons of napalm were dropped. These bombing raids were looked upon as preparatory for later air attacks on the Japanese homeland. On this basis, these missions were used to experiment with the effect of different types of bombs and fuses, as well as tactics best suited to achieving various objectives[17].

9.15 The impact on Japan's fuel supplies

In 1937, the air arm of the Japanese fleet was using 85-octane aviation fuel. By 1939 this had been increased to 87 octane, along with a basic 76-octane fuel for training. By 1942, 92-octane combat aviation fuel was in general use, but in 1943 this was lowered to 91 octane, while permissible lead content and boiling temperature were increased. This grade of fuel was used for combat for the remainder of war, along with 89-octane N-1 naval sub-standard fuel, until the last months when the octane number was dropped to 87 for summer grade fuel. Analysis of Japanese aviation gasoline undertaken by the US, both during and after the war, confirmed that the majority of the samples tested had an octane rating of between 87 and 93. Very limited quantities of 95 and 100 octane gasoline were only produced during the late stage of the war.

Records show that, except for brief periods in early 1942, fuel consumption by Japan exceeded production and imports. By the middle of 1943 US submarines were so successful at sinking tankers that the Japanese Navy had already begun to be affected by the shortage of fuel. Initially, training cruises were shortened, and then eliminated. Strategic decisions had to be made based on fuel requirements rather than on political or military imperatives. Aviation fuel existed in sufficient quantities for both the Japanese Army and Navy until around mid 1944, before shortages began to have a serious impact. However, from late 1943 flight commanders had already begun to instruct pilots on how to conserve fuel. When the shortage of aviation fuel began to become significant, the immediate effect was on pilot training. As with all air forces, training accounted for by far the greater part of all fuel consumption.

From the middle of 1944, when the aviation fuel situation became serious, the Japanese began substituting standard aviation fuel with a mixture of gasoline and ethanol called A-Go. The use of ethanol as a blending agent in aviation gasoline was impeded by the insolubility characteristics of 95% ethanol, and the lack of equipment in Japan for manufacturing anhydrous ethanol. The use of ethanol involved no loss in power but caused numerous other problems. These included poor acceleration, an approximate 50% increase in the rate of fuel consumption, inadequate combustion that caused cylinder super-cooling, excessive corrosion of parts of the fuel system, faulty distribution of air fuel mixture, starting difficulties in cold weather and separation of the ethanol-gasoline mixture. The use of ethanol, alone and in mixtures of 20% and 50% gasoline, came into general use for training purposes by both the Army and Navy early in 1945. In July 1945 ethanol comprised about 21% of the Army Air Force consumption of aviation fuel, and about 40% of the Navy's consumption. The A-Go fuel proved to be dangerously unreliable, causing engines to cut out in flight, resulting in numbers of trainee pilots being killed. Due to the shortage of fuel, army trainees resorted to flying gliders during the first month of training. Ethanol was also used extensively as motor fuel. In 1941 motor fuel consisted of 90% gasoline and 10% ethanol. Later gasoline was reduced to 80% and ethanol was increased to 20%. Towards the end of the war 100% ethanol was used as motor fuel for military trucks, along with acetone and methanol.

Fuel shortages began to affect all combat operations by mid-1944, at a time when US air activity was reaching its peak. As a result the Japanese were forced to cut down consumption during the very period when the air war was reaching its height. Consumption dropped from an average of 503,000 barrels a month during the period April to June 1944 to 142,000 barrels during July and August of 1945. The shortage of aviation fuel is considered to have been an important factor in the decision of the Japanese High Command to create a suicide kamikaze force. It was considered that three suicide planes were sufficient to sink an American warship, whereas a conventional attack required 15–20 aircraft. In addition, fuel was saved because no fuel was required to return to base. Towards the end of the war the lack of fuel had made it extremely difficult for the Japanese to continue to train pilots, but they still hadn't resorted to using ethanol for combat operations. They were planning to do so if excessive depletion of fuel stocks made such an expedient necessary.

By the end of 1944, the stocks of around 4,300,000 barrels of oil that the Japanese had at the time of Pearl Harbour had been reduced to just over 1,000,000 barrels. Civilians were recruited in an attempt to brew fuel from pine roots. This futile effort would have required more than a million workers to meet the target. Besides, the pine root fuel gummed up engines beyond repair after only a short running time. Towards the end of the war the Japanese also attempted to revive their synthetic fuel process, and entered into an agreement with I. G. Farben to provide technical assistance. However, the defeat of Germany brought this endeavour to an end. By 1945 the USAAF had over 1,000 Superfortress B-29 heavy bombers in service. Escorted by P-51 Mustang fighters, the B-29s roamed the skies practically unchallenged, bombing virtually every city in Japan. City after city was incinerated by deadly napalm attacks, thermite bombs and fragmentation bombs. By March 1945, most of the Japanese aircraft factories had been damaged or destroyed and only 500 fighters were available for the air-defence of Japan. This limitation was exacerbated by a severe shortage of pilots and training instructors. Precedence was given to conserving aircraft, pilots and fuel for defence against invasion. This led to a decision to cut down on the interception of B-29 bombers, and to employ ramming tactics. Although oil imports had ceased, and Japanese refinery outputs had been reduced to around 6% of previous levels, Japan still managed to produce about one-fourth of the aviation fuel consumed by the Japanese air forces during July and August 1945. When the war ended Japan still possessed aviation fuel stocks of slightly over 1,000,000 barrels. This would have been sufficient for about seven month's supply for suicide operations, or for about two month's supply for orthodox air operations, at the rate of the greatest previous consumption. Inasmuch as the principal air defence planned to repel an allied invasion was to take the form of suicide attacks against surface vessels, the Japanese would have run out of aircraft well before they ran out of fuel, had an invasion become necessary.

9.16 Post-war developments in Taiwan leading to the re-establishment of the butanol fermentation process

After the war ended, the Republic of China, led by the Kuomintang, became responsible for governing Taiwan[45,46]. In 1945, soon after the Japanese surrender, Chiang Kai-shek appointed Chen as the first governor-general of the Taiwan Provincial Administration Executive. During the next two years, he nationalised all physical assets on Taiwan to do with industry, trade, commerce and transportation. He also confiscated all Japanese and Japanese–Taiwanese owned lands, industries and business assets, and transferred these to the new Nationalist state enterprises. The four largest private sugar companies under Japanese control were amalgamated into the Taiwan Sugar Company, and Taiwan's six oil companies were merged into the China Petroleum Corporation. The new administration also set up two new overarching control agencies, the Trade Bureau and the Monopoly Bureau, which controlled the import and export of all major goods and raw materials, and oversaw the production and marketing of foodstuffs and manufactured products. By the end of 1946, Taiwan's provincial government had taken control of most of the economy and industrial wealth, but Chen's central command system was unable to revitalise the island's economy. By this time the new administration was facing increasing challenges as inflation escalated to hyperinflation, unemployment skyrocketed and the political rift between the mainlanders and the pre-1945 Taiwanese continued to widen. By the end of 1946 Taiwanese resentment toward the provincial administration boiled over, resulting in a revolt that erupted in February 1947. The uprising lasted for a week, during which the Taiwanese gained control. A large force of Chinese troops were dispatched from the mainland, and the uprising was brutally suppressed. Damage was extensive, and the death toll was estimated to be around 20,000. As a result a series of political and economic reforms were implemented[45,46].

Kagi, which was renamed Chiayi after the war, had been subjected to extensive bombing, and much of the town had been destroyed. The majority of the population were reduced to living in makeshift shacks made from salvaged material, and an enormous amount of clearing up was required. Ninety percent of the butanol plant had been destroyed, and fire had added to the damage. Unemployment was a serious problem. Farmers in the surrounding area were also in a difficult position, as much of the surrounding countryside had been devoted to growing sweet potatoes for the factory. The authorities were apparently indifferent to the distress of the townspeople, and adopted a cautious position regarding the future of the factory. One problem was that most of the Japanese had been repatriated to Japan after the war, so much of the technical expertise, knowledge and access to spare parts had been lost. As little money was available, the number of people employed at the factory was reduced to about 130, and they were unable to carry out even the most basic repairs necessary to keep what remained of the plant from deteriorating. This 'wait-and-see' policy rankled with the townspeople, who were also suffering because lumber, the other major industry of the town, was hampered by a shortage of machinery. Dr Hou was the engineer in charge of the butanol factory, and it is apparent that he undertook a number of visits to Shanghi to try and raise capital needed to rebuild the factory with the intention of producing butanol, isopropanol, acetone and ethanol by fermentation[34,46]. These efforts met with success and rebuilding of the plant on the original factory was started in 1946 (Fig. 9.8). This resulted in the establishment of the Chiayi Solvent Works in May 1947. The first test runs were undertaken in September 1947. By the end of December, good quality butanol, meeting international standards, was being produced and the factory went into full production early in 1947. During the first few years of production both sweet potatoes and cassava were used as the substrate, producing solvent in the ratio of around 56% butanol, 30% acetone and 14% ethanol. The export of butanol to Britain was begun in 1948. These shipments, which went to Imperial Chemical Industries, were the first chemical exports from Taiwan. When the Chinese Communists finally won the Civil War in 1949, the government of Chiang Kai-shek, along with 1,500,000 Chinese who supported the Kuomintang, left for Taiwan, and joined the 6,000,000 Taiwanese already residents there[45]. The Chinese Petroleum Corporation (CPC), which had been formed in Shanghai in 1946 also moved its headquarters to Taiwan in 1949[47]. The CPC was charged with the task of developing oil refining facilities, supplying energy and

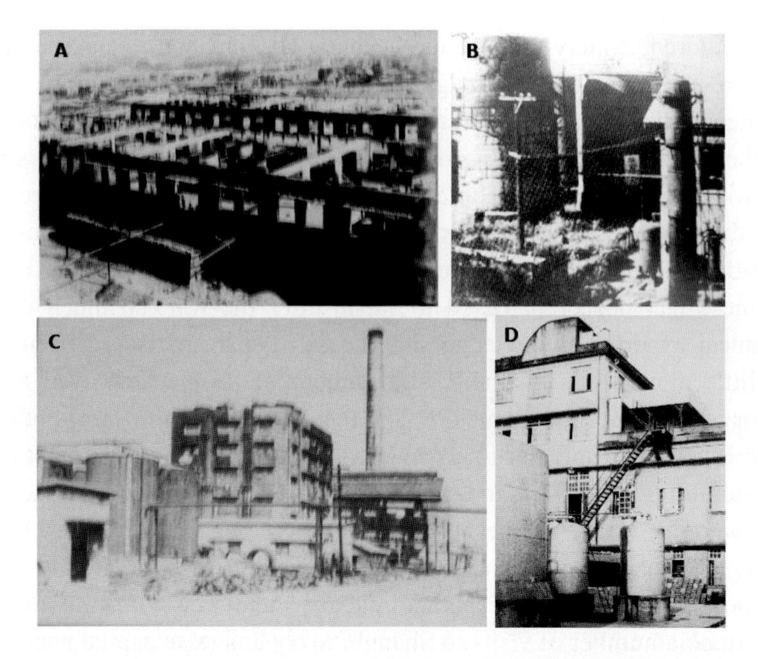

Fig. 9.8: The butanol plant at Chiayi (Kagi) after WWII. A. and B. Rebuilding the damaged fermentation plant in 1946. C. and D. The newly rebuilt plant in 1947.

promoting the petrochemical industry, and the Chiayi Solvent Works was placed under its control.

The increasing price of raw materials for the fermentation process resulted in a gradual switch from sweet potato to cassava, and then to molasses. In 1954 the factory shifted to using wheat starch for two years. In 1958 the ABE fermentation was stopped for a time due to the high cost of raw materials. The process was restarted for a period in 1959, but was finally abandoned after having operated for a period of 19 years. The fermentation plant was subsequently demolished and the building refurbished for other uses. By 1958 the CPC had expanded into the petrochemical industry, and in 1959 the first aromatic solvent unit was erected at the factory. The production of benzene, toluene and xylene began at the plant in 1960. These were used to make products such as naphtha and rubber solvents to supply the needs of the domestic market. A new naphtha cracking plant was put into operation in 1968[47]. The CPC chemical plant is still in operation on the original 1939 factory site, with the basic layout largely unchanged (Fig. 9.2). This site now also houses the CPC's main research facility.

The fermentation process at the plant in Chiayi utilised Japanese strains during the war. After the war the plant continued with two main strains designated the C and the T strains. CPC reports in 1950 and 1951 describe optimisation studies, and contain descriptions of the morphological, cultural and physiological characteristics of two strains of *Clostridium* used for the fermentation of cane molasses[48,49]. A later report mentions that the butanol fermentation process using molasses was being carried out at both pilot-plant and industrial-plant scale. Another report in 1954 mentions that cassava was an ideal raw material for butanol and ethanol fermentation because its starch content was significantly higher than that of sweet potato. Articles published by Petroleum Communications of the CPC included optimisation of the butanol fermentation, and trends in world solvent production over the previous ten years, along with new developments and management of the Chiayi Solvents Works[34,50,51]. In 1956 the Food Technology Division of the Joint Commission on Rural Reconstruction introduced bi-weekly forums dealing with current problems of Taiwan agriculture. Topics included the butanol fermentation and sugar refining. In 1958 a Japanese patent (244,905. Publication No. Sho. 33-3798/1958), covering the process, was issued to the company.

During the period from 1952 to 1954 Baba published a series of seven papers relating to the butanol iso-propanol fermentation process in Taiwan[52,53,54,55,56,57,58]. A summary of these findings is available in English[3]. These reports covered a number of aspects of the industrial production of butanol, isopropanol and acetone using *C. toanum,* including studies on the course of solvent formation and gas production in the batch process. The information contained in these reports indicates that the raw materials used as the fermentation substrate included blackstrap molasses, sugar cane juice, sugar syrup and raw sugar. Blackstrap molasses was a major fermentation substrate, and the influence of different types of molasses was investigated. Molasses produced from the clarification of raw sugar juice by the defecation process was satisfactory, whereas that produced by the clarification using the carbonation process was found to be unsatisfactory. The fermentable sugar concentrations used varied between 7.0 and 7.3%, which gave solvent yields of 30–33%. The standard additives to the mash included 3% ammonium sulphate, 0.3% rice bran and 0.3% calcium carbonate. The batch process was maintained at temperatures of between 33 and 37°C, and the fermentation was completed after 30–40 hours. The titratable acids reached a maximum after 10–15 hours, and declined rapidly after 20–25 hours. The solvent ratios were strongly influenced by the nature of the fermentation substrate, and ratios varied between 53 and 65% butanol, 19 and 44% iso-propanol, and 1 and 24% acetone. The proportion of acetone decreased as the concentration of isopropanol increased. Solvents were separated and purified by continuous distillation. The weight of gas produced was more than 1.5 times that of the solvents produced. Baba also undertook research on solvent toxicity. He reported that the addition of 0.5% butanol to sugar-based mash 15 hours after inoculation arrested solvent production. The addition of 1% butanol almost completely suppressed the fermentation. Acetone, added at a concentration of 1%, had a slight inhibitory effect on fermentation activity, whereas 1.5% acetone resulted in a marked reduction of total solvent yield. Addition of 0.9% iso-propanol, which was the level normally produced in the fermentation, only had a slight inhibitory effect. The addition of 1.5% iso-propanol markedly reduced the yield of total solvents.

9.17 The post-war butanol fermentation process in Japan and China

Following WWII, the US occupation authorities forced many Japanese businesses to undergo restructuring as a result of their wartime activities. Amongst these was the Kyowa Hakko company, which had played a key role in the development of the butanol fermentation for the production of aircraft fuel as part of Japan's war effort. The Kyowa Fermentation Industry Company resumed commercial activities and, in 1948, they established the first industrial ABE fermentation plant in Japan for the production of solvents using molasses or sugar syrup as the substrate[30]. The factory was established at the company's plant in Yokkaichi and operated for about 14 years, from 1948 until around 1961, when the fermentation process was superseded by a synthetic process for the production of solvents. Little information regarding the details of this plant or its operation appear to have been released. It can be concluded that phage infections remained a problem, as a number of applied studies were carried out on phage infection, as well as fermentation optimisation, by the company at the Tokyo Research Laboratory. These studies resulted in a series of publications in Japanese[59,60,61,62].

At least one other Japanese company operated a commercial ABE fermentation process. The Sanraku Distiller's Company established a plant at their Yatsushiro plant in the Kumamoto prefecture some time during the 1950s. Again very little information regarding the details of this plant or its operation appear to have been published. In an endeavour to increase the ratio of butanol produced in the fermentation, Hongo and Nagata isolated a new high butanol-producing *Clostridium* from soil in 1959 and named it *Clostridium saccharoperbutylacetonicum*. This strain was patented[3] and subsequently proved to be a new species of solvent-producing *Clostridium*[39]. The new isolate proved to be very efficient, but was found to be susceptible to phage infection. Phage contamination occurred twelve times during one year[64,65]. These phages fell into three morphological and serological groups, and these setbacks appeared to be partly responsible for process being abandoned in the early 1960s[66]. Hongo, Ogata and their colleagues continued to study these phages and associated clostocins for most of the next two decades, and published numerous papers in this area. A 1963 worldwide survey of fermentation industries, prepared for the International Union of Pure and Applied Chemistry, reported that in 1963 Japan's annual production of acetone-butanol by fermentation was around 15,000 tons.

The industrial ABE fermentation was started in China in the early 1950s at the Shanghai Solvent Plant, established in Pudong, Shanghai, to produce solvents by batch fermentation. It is known that personnel from the butanol factory in Taiwan visited Shanghi during the late 1940s, and it can be speculated that this contact may have had some bearing on the establishment of the ABE fermentation process in China. Due to the rapid expansion of the chemical industry and increased penicillin production, a study to boost the national ABE industry was initiated in 1958. A recent review by Chiao and Sun[14] provides a comprehensive history of the subsequent expansion of the ABE fermentation industry in China along with the development of continuous production technology.

Acknowledgement

The valuable contributions made to this chapter by the following are gratefully acknowledged. Professor Ralph Kirby for arrangements and support for a visit to Taiwan. Dr Huang and Dr Chen for hosting a visit to the CPC Company factory site in Chiayi and the provision of photographic and other historical material. Professor Carton Chen for arrangements and the translation of Chinese documents. The Oenon company for providing historical documents. Dr Yukinori Iwasaki for the translation of the Japanese documents. Dr John Hauman and Wyn Jones for editorial input.

References

1. Gabriel C. L. and Crawford F. M., *Ind. Eng. Chem.* 22 (1930).
2. Jones D. T. and Woods D. R., *Microbiol Rev.* 50 (1986).
3. Prescott S. G. and Dunn C. G., *Industrial Microbiology*, 3rd Edition (McGraw-Hill Book Co., New York, 1959).
4. Dürre P. and Bahl H., in *Products of Primary Metabolism*, 2nd Edition, Roehr M., Ed. (VCH Publisher, Weinheim, Germany, 1996), pp. 229–268.
5. Rogers P. *et al.*, in. *Prokaryotes 1*, Dworkin M. *et al.*, Eds (Springer, Berlin, 2006), pp. 672–775.
6. Beesch S. C., *Eng. Proc. Dev.* 44 (1952).
7. Beesch S. C., *Appl. Microbiol.* 1 (1953).
8. McCutchan W. N. and Hickey R. J., *Ind. Ferment.* 1 (1954).
9. Ryden R., in *Biochemical engineering*, Steel R., Ed. (Heywood, London, 1958), pp. 125–148.
10. Hastings J. H. L., in *Economic microbiology,* Vol. 2, Rose A. H., Ed. (Academic Press, Inc., New York, 1978), pp. 31–45.
11. Spivey M. J., *Process Biochem.* 13 (1978).
12. Jones D. T., in *Clostridia: Biotechnology and Medical Applications*, Bahl H. and Dürre P., Eds (Wiley, VCH, Weinheim, 2000), pp. 125–166.
13. Zverlov V. V. *et al.*, *Appl. Microbiol. Biotechnol.* 71 (2006).
14. Chiao J. S. and. Sun Z. H., J. *Mol. Microbiol. Biotechnol.* 13 (2007).
15. Maddo R. F., *West Virginia History Journal* 55 (1996). Retrieved January 2011 from www.wvculture.org/history/journal_wvh/wvh55-6.htm.

16. Antonucci M., *Command Magazine* 20 34–40 (1993).
17. Craven W. F. and Cate J. L. (Eds.), *The Army Air Forces in World War II*, Vol. V (University of Chicago Press, Chicago, 1953).
18. Maurer I., *U.S. Govt. Print. Off., Supt.* DOCSN. O.:D 301.2:C73/3/983 D790.A533 (1983).
19. Wolborsky S. I., Thesis, School of Advanced Airpower Studies, Air University, Albama (1994). Retrieved January 2011 from www.dtic.mil/cgi-bin/GetTRDoc?Location=U2&doc=GetTRDoc.
20. Reports X-38 (N) 1 to 10 (2011). Retrieved January 2011 from http://www.fischer-tropsch.org/primary_documents/gvt_reports/USNAVY/USNTMJ%20Reports/USNTMJ_toc.htm.
21. *Combat Chronology of the US Army Air Forces*. (n.d.). Retrieved January 2011 from ftp.rutgers.edu/~mcgrew/wwii/usaf/html/Jun.45.html.
22. *HyperWar: The Army Air Forces in WWII: Vol. V-The Pacific* (n.d.). Retrieved January 2011 from www.ibiblio.org/hyperwar/AAF/V/AAF-V-16.htm.
23. *U.S. Air force Historical Research (AFHRA) acquisition achiements* (n.d.). Retrieved January 2011 from http://gis.rchss.sinica.edu.tw/GIArchive/?page_id=438.
24. *U.S. Strategic Bombing Survey (Pacific): Japanese Air Target Analyses* (n.d.). Retrieved January 2011 from www.footnote.com/document/26164985/.
25. *Targets in area of Formosa (Taiwan)* (n.d.). Retrieved January 2011 from gis.rchss.sinica.edu.tw/GIArchive/wp-content/.../02/Catalog_TPPIR_1.pdf.
26. *WWII Formosa Japan Oil Fields* (1945) (2011). Retrieved January 2011 from www.scribd.com/doc/43228345/WWII-Formosa-Japan-Oil-Fields.
27. Palucka T., *Invention and Technology Magazine* 20 (2005).
28. Kawabe N., *Overseas Activities and their Organization*, Chapter 7 (2011). Retrieved January 2011 from d-arch.ide.go.jp/je_archive/pdf/book/jes3_d08.pdf.
29. Rokusho B. and Yamazaki M., *Agr. Biol. Chem.* 33 (1969).
30. *Kyowa Hakko Kogyo Co., Ltd. Company History* (n.d.). Retrieved January 2011 from http://www.fundinguniverse.com/company-histories/kyowa-hakko-kogyo-coltd-history/.
31. Kubo F. M., *A reorganization of the modern sugar manufacturing industry* (2007). Retrieved January 2011 from www.21coe-win-cls.org/8b/20080126/paper/Kubo_final.pdf.
32. *Taipei Imperial University Historical Report: Opening of the Southern Research Institutions* (n.d.).
33. Show I., *World Wide Chemistry* (1939). Retrieved January 2011 from pubs.acs.org/doi/abs/10.1021/cen-v017n008.
34. Ye C. S., *Petroleum Communications. China Petroleum Corporation. Taiwan, China* 54 (1955).
35. *Time Mag news report* (1940). Retrieved January 2011 from www.ww2f.com/war-pacific/22718-japanese-java-dec-1940-a.html.

36. Baba T., *J. Agric. Chem. Soc. Japan*. 19 (1943).
37. Asai T. and Haruda S., *J. Agric. Chem. Soc. Jap*. 43 (1943).
38. Breed S. *et al.,* Eds, *Bergey's manual of determinative bacteriology.* 7th Edition (The Williams & Wilkins Company, Baltimore, 1957).
39. Keis S. *et al.*, *Inter. J. Syst. Evol. Microbiol*. 51 (2001).
40. Shoutens G. H. and Groot W. J., *Process Biochem*. 20 (1985).
41. van der Eng P., *Food Supply in Java during War and Decolonization*, 1940–1950 (2008). Retrieved January 2011 from http://mpra.ub.uni-muenchen.de/8852/MPRA Paper No. 8852.
42. *Military History of the Philippines during World War II.* (n.d.). Retrieved January 2011 from en.wikipedia.org/.../Military_history_of_the_Philippines_during_World_War_II.
43. Umemura T. *et al.,*. *Studies on butanol fermentation: Research period 1941–1944* (1944). Prepared for US Naval Technical Mission to Japan — Report X-38(N)-3: Enclosure (B) 5.
44. Ogata S. and Hongo M., *Adv. Appl. Microbiol*. 25 (1979).
45. Kerr G. H., *Formosa Betrayed* (Houghton Mifflin, Boston, 1965).
46. Shackleton A. J., *Formosa Calling* (Taiwan Publishing Company, Taiwan, 1998).
47. *Chinese Petroleum Corporation — Company History.* (n.d.). Retrieved January 2011 from www.fundinguniverse.com/.../Chinese-Petroleum-Corporation-Company. History.html.
48. Bah H. S. *et al.*, *Rep. Taiwan Sug. Exp. Sta*. 6 (1950).
49. Bah H. S. *et al.*, *Rep. Taiwan Sug. Exp. Sta*. 7 (1951).
50. *Petroleum Communications, China Petroleum Corporation, Taiwan, China.* 60 (1956).
51. *Petroleum Communications, China Petroleum Corporation, Taiwan, China* (1958).
52. Baba T., *Bull. Fac. Eng. Hiroshima Univ*. 1 (1952).
53. Baba T., *Bull. Fac. Eng. Hiroshima Univ*. 2 (1953a).
54. Baba T., *Fac. Eng. Hiroshima Univ*. 2 (1953b).
55. Baba T., *Fac. Eng. Hiroshima Univ*. 2 (1953c).
56. Baba T., *Fac. Eng. Hiroshima Univ.* 3 (1954a).
57. Baba T., *Fac. Eng. Hiroshima Univ*. 3 (1954b).
58. Baba T., *Fac. Eng. Hiroshima Univ*. 3 (1954c).
59. Kinoshita S. and. Teramoto K., *Nippon. Nogei. Kagaku. Kaishi*. 29 (1955a).
60. Kinoshita S. and. Teramoto K., *Nippon. Nogei. Kagaku. Kaishi*. 29 (1955b).
61. Kinoshita S. and. Teramoto K., *Nippon. Nogei. Kagaku. Kaishi*. 29 (1955c).
62. Kinoshita S. and. Teramoto K., *Nippon. Nogei. Kagaku. Kaishi*. 29 (1955d).
63. Hongo M., *Butanol by fermentation*. U.S. patent 2,945,786. (1960).
64. Hongo M *et al.*, *J. Agric. Chem. Soc. Jpn* 39 (1965).
65. Hongo M *et al.*, *J. Agric. Chem. Soc. Jpn* 39 (1965).
66. Hongo M. and. Murata A., *J. Ferment. Association Jpn* 24 (1966).

INDEX

A

B

C

D

E

F

G

H

I

L

M

N

O

P

Q

R

S

T

U

V

Y